D1548907

DARWIN'S AUDUBON

SCIENCE AND THE LIBERAL IMAGINATION

New and Selected Essays by
Gerald Weissmann

ALSO BY GERALD WEISSMANN

THE WOODS HOLE CANTATA

THEY ALL LAUGHED AT CHRISTOPHER COLUMBUS

THE DOCTOR WITH TWO HEADS

THE DOCTOR DILEMMA

DEMOCRACY AND DNA

YEAR OF THE GENOME

DARWIN'S AUDUBON

SCIENCE AND THE LIBERAL IMAGINATION

New and Selected Essays by
Gerald Weissmann

PERSEUS
PUBLISHING

Cataloging-in-Publication Data available from the Library of Congress

ISBN 0-7382-0597-4

Perseus Publishing is a member of the Perseus Books Group

First paperback printing, December 2001
1 2 3 4 5 6 7 8 9 10—05 04 03 02 01

Perseus Publishing books are available at special discounts for bulk purchases in the U.S. by corporations, institutions, and other organizations. For more information, please contact the Special Markets Department at the Perseus Books Group, 11 Cambridge Center, Cambridge, MA 02142, or call (800) 255-1514 or (617) 252-5298, or email j.mccrary @perseusbooks.com.

Visit us on the World Wide Web at http://www.perseuspublishing.com

FOR MY GRANDSON, BENJAMIN

Through each new sense the keen emotions dart
Flush the young cheek and swell the throbbing heart.
From pain and pleasure quick VOLITIONS rise
Lift the strong arm or point the inquiring eyes;
With reason's light bewilder'd Man direct,
And right and wrong with balance nice detect.
Last in thick swarm ASSOCIATIONS spring.
Thoughts join to thoughts, to motions motions cling:
Whence in the long trains of catenations flow
Imagined joy and voluntary woe.

—Erasmus Darwin, from *Temple of Nature,* Volume I

CONTENTS

Introduction   1

1 Darwin's Audubon   9

2 Inflammation as Cultural History   25

3 Gertrude Stein and the Ctenophore   35

4 Puerperal Priority   49

5 Bête Noire   61

6 Science Fictions   69

7 Call Me Madame   81

8 The Woods Hole Cantata   87

9 Foucault and the Bag Lady   97

10 In Quest of Fleck: Science from the Holocaust   107

11 Auden and the Liposome   115

12 No Ideas but in Things   125

13 Nobel Week 1982   137

14 They All Laughed at Christopher Columbus   149

15 Westward the Course of Empire   159

16 *Nullius in Verba*: Lupus at the Royal Society   179

17 Springtime for Pernkopf   191

18 The Baron of Bellevue   207

19 The Doctor with Two Heads   215

20 Wordsworth at the Barbican   235

21 Losing a MASH   245

22 To the Nobska Lighthouse   255

23 Titanic and Leviathan   269

24 Daumier and the Deer Tick   283

    Sources   311

    Credits   325

    Index   329

    Acknowledgments   339

INTRODUCTION

There is no form of lead-poisoning which more rapidly and thoroughly
pervades the blood and bones and marrow than that which reaches the
young author through mental contact with type-metal. Qui a bu,
boira,—he who has once been a drinker will drink again, says the French
proverb. So the man or woman who has tasted type is sure to return to
his old indulgence sooner or later. In that fatal year [1830; "Old
Ironsides"] I had my first attack of authors' lead-poisoning, and I have
never got quite rid of it from that day to this.

O. W. HOLMES, *Medical Essays* (1884)

I'm afraid that the essays in this book are a flagrant symptom of lead poisoning, acquired as in Holmes' case, very early in life. The disease began in my freshman year on Morningside Heights and I can trace the contagion directly to Mark Van Doren. He stood by his desk and stared out over our heads through the window at the frieze of great names carved into the facade of Butler Library. Harris tweeds, elfin smile, and that shock of hair! One foot was perched on the seat of an empty chair, his elbow was propped on his knee. Chin in hand, he nibbled a piece of chalk, looked up at the sky, then turned to us and wondered—apropos of Kafka at the time—"I wonder if we can call any imaginary act good or bad?" I was not sure what he meant in the spring of 1948, and I'm still not sure, but I remember thinking that I had watched him turn from a meditation on his colleagues, for it was not only as a teacher but as a writer that Van Doren read the works of Dante, Milton, and Shakespeare—those names carved into the stone of Butler Library. I knew then and there that one could do far worse than to spend one's life at a university where one was expected not only to teach well but to write something that might find its way to a library.

As luck would have it, many of the essays I have chosen for this book have indeed found their way to the libraries. Together with more recent works, they constitute what the art world would call a retrospective. But since they represent a three decade-long effort to bridge the two cultures, they are also a memoir of parallel loves. Chekhov insisted that while he remained married to medicine, literature was his mistress; in this context I confess to bigamy. Wilde, of course, had the last word on the matter: Bigamy, he wrote, is having one wife too many. Monogamy is the same.

Be that as it may, the social views these essays take may seem quaint compared to the trendy tilts of our new Unreason. While it's a fair bet that our century will be remembered for having cracked the genetic code, landed on the moon, and toppled agelong prejudice, much of its high art and higher thought—from Yeats to Heidegger, from Eliot to Foucault—have been crafted in rage against twentieth-century science. Pop thought has followed suit. Despite or perhaps because we've broken the sound barrier and doubled our life span in a century, our few remaining bookstores are jammed with millennial fantasies, with platitudes plucked from the East or mud-cures distilled from mildew. Well, we've been down that path before. As Robert Darnton has taught us, the first age of modern science—the age of Buffon, Franklin, and Lavoisier—dissolved into highfalutin nonsense when the intelligentsia turned to the likes of mad Marat, or Mesmer, or Swedenborg. No matter, two centuries later H_2O still dissociates to H^+ and OH^-, electricity revives the faltering heart, and O_2 keeps it pink and beating. *Beneficentiam et veracitatem est* [it is beneficial and verifiable], wrote Longinus of the sublime, and it is as true of medical science as of the *Nike of Samothrace*. Both have outlasted magic and Mesmer, and in these essays I've tried to explain why.

In the eighteenth century science struck back at Mesmer and Ben Franklin took the lead in that effort. The fastidious ministers of the *ancien régime* had heard enough of Mesmer and his disciples, of those Mesmeric touchy-feely sessions in which "vapors and fluids" were exchanged. They appointed a commission to determine whether Mesmerism "worked" in practice. The group consisted of four prominent doctors from the faculty of medicine, including Guillotin (of the blade), and five members of the Academy of Sciences, including Bailly (of Jupiter), Lavoisier (of oxygen), and Benjamin Franklin (of the spark). Years later, Oliver Wendell Holmes ranked Franklin with the Montgolfiers (of the balloon) and Morton (of anesthesia):

We've tried reform—and chloroform—and both have turned our brain;
When France called up the photograph, we roused the foe to pain;
Just so those earlier sages shared the chaplet of renown,—
Hers sent a bladder to the clouds, ours brought their lightning down . . .

Those earlier sages, the commissioners, spent weeks listening to Mesmeric theory and observing how its patients fell into fits and trances. They underwent continuous mesmerizing themselves, with no effect, and then tested the operation of Mesmeric "fluid" outside the excitable atmosphere of the Mesmeric clinic. They found false a report that being mesmerized through a door caused a woman patient to have a crisis. Another "sensitive" patient was led up to each of five trees, one of which had been mesmerized, in Franklin's garden at Passy; he fainted at the foot of the wrong one. Four normal cups of water were held before a mesmerized patient at Lavoisier's house; the fourth cup produced convulsions, yet she calmly swallowed the mesmerized contents of a fifth cup, which she believed to be plain water. The commissioners concluded that the effects of mesmerizing could be attributed to the overheated imaginations of the mesmerists.

Many of the essays in this book advance the argument that scientific reasoning, like Enlightenment thought itself, has a strong aesthetic component, that science has as much to do with form as with function. To test this notion one has only to compare David's portrait of the gentle Lavoisiers with his image of a tousled Marat or a *Life* photograph of white-clad Lewis Thomas with its portraits of our hirsute naturopaths.

No one has championed the aesthetic, formal aspect of science better than the founder of modern biology, Georges-Louis Leclerc, Comte de Buffon (1707–1788). Buffon's *Histoire Naturelle* was to biology what Diderot's *Encyclopédie* was to the general knowledge of the time. Indeed, Buffon wrote several articles for the *Encyclopédie* and, as patron and protector, was a character witness when Diderot was threatened with imprisonment for impiety. Buffon was one of those Enlightenment polymaths, at once a stylist, a member of the Academy, and director of the Royal Garden (today's Jardin des Plantes with its Musée d'Histoire Naturelle). His work demonstrates that the epigram is to prose what the equation is to science, a phase change from the amorphous to the crystalline. But Buffon had it first; the subject of his inaugural speech at the Royal Academy was not science, but style:

Bien écrire, c'est tout à la fois bien pensée, bien sentir et bien rendre. C'est avoir en même temps de'espirit, de'lâme et du gout.
[To write well is at once to think, feel and express oneself well; simultaneously to possess wit, soul and taste.]

With disdain for fanciful systems based on "natural law," he noted that

Il est plus aisé d'imager un système que de donner une thêorie.
[It is easier to dream up a system than to work out a theory.]

Buffon had three younger associates at the Jardin de Plantes: Geoffroy Sainte-Hilaire, Georges Cuvier, and the chevalier Lamarck. These were the men who hammered out the French version of evolutionary theory; their politics were as pointedly secular and Whiggish as those of the Darwins on the other side of the channel. When Lamarck assured his readers in 1809 that *Dans tout a que la nature opére, elle ne fait rien brusquement* [Everywhere nature is at work, she does nothing abruptly] his compass was not limited to zoology. After Marat and the Terror, cataclysms had little appeal. In like vein, my own interest in the two cultures favors the gradual over the sudden, sure change over blanket upheaval. That position, in accord with Enlightenment teaching, can best be described as Whig, a term that has the convenience, as Lord Russell put it, of expressing in one syllable what Conservative Liberal expresses in seven.

The essays in this book are in debt to notions made popular by latter-day Whigs: the liberals, meliorists, and Fabians. These range from George Eliot and George Bernard Shaw to Jaques Barzun and—especially—Lionel Trilling, who first introduced me to The Liberal Imagination. Like eighteenth-century Whigs, these liberal skeptics based their hopes for a rational future on the bedrock of practical science. That line of belief in progress through reason makes the Whigs, meliorists, and Fabians the intellectual ancestors of today's molecular biology. Their skeptical views extended to the field of general belief as summed up by Wilfred Trotter in 1939:

The various systems of doctrine that have held dominion over man have been demonstrated to be true beyond all question by rationalists of such power—to name only the greatest—as Aquinas and Calvin and Hegel and Marx. Guided by these master hands the intellect has shown itself more deadly than cholera or bubonic plague and far more cruel. The incompatibility with one another of all the great systems of doctrine might surely have been expected to provoke some curiosity about their nature.

The Fabians are remembered nowadays chiefly for their publicists, H. G. Wells and George Bernard Shaw, rather than for the activist founders of the Society, Beatrice and Sidney Webb. The Society took its name from the Roman general Fabius Maximus, whose tactics in fighting Hannibal in the second Punic war earned him the nickname "the delayer," since he avoided pitched battles to nibble away at his enemy's supply lines and badger him by skirmishes. Fabians eschewed direct battle with established government, working, instead, through public education, the journals, and the lecture platform. Although the Fabians were tough on religion, they were soft on the pious. They cooperated with the meliorist churchmen who began the settlement house movement in London's East End. In America, Jane Addams and her friends patterned Hull House and the university settlements on the London example. Also based on a military model, the Salvation Army was the evangelical counterpart of the Fabians at the vanguard of active philanthropy. No wonder, then, that Shaw's character Major Barbara was his homage to the meliorist militant. But Shaw knew that nothing useful would come of all that do-goodery without attention to practical science. The real hero of *Major Barbara* is Andrew Undershaft, a clear-eyed munitions manufacturer, Shaw's portrait of the meliorist, techno/commercial.

In retrospect, I realize that from the very first ("The Woods Hole Cantata"), my essays have carried on the debate between Barbara and Undershaft, between creed and deed, art and science, form and function. By and large, I tilt towards Undershaft, believing with him that the industrial application of the second law of thermodynamics, i.e., $S > q/T$, will do more to erase the "eighth deadly sin, poverty" than all the good works in the world. And if defining entropy implies accepting its dangers, so be it. As Whitehead wrote in his Fabian mode, "It is the business of the future to be dangerous; and it is among the merits of science that it equips the future for its duties."

Years ago, a Cambridge don spun a Shavian anecdote that makes the point. It seems that in the depth of the Great Depression, Louis B. Mayer, the Hollywood mogul, cabled an offer to Shaw to come to California to discuss the bringing of his classics to the screen—*Major Barbara, Pygmalion,*

The Doctor's Dilemma, and others. The terms of the deal were so staggering
that the octogenarian Shaw was persuaded to abandon his cozy nest in Ayot
St. Lawrence, which he rarely left, and chug across the water to a land which
he held in little esteem. Shaw packed a skimpy Gladstone and grumped
throughout the *Queen Mary's* first-class formal dinners in his worn country
tweeds and Norfolk jacket. After six vegetarian days at sea, he disembarked
at New York only long enough to curl up on the fashionable Twentieth
Century train heading west. Two more days and the dyspeptic dramatist
arrived in L.A. where the perpetual sun and a chauffeured Packard limousine awaited him.

Shaw was whisked past studio guards, gates, sets, starlets, flunkies,
stairways, and antechambers to meet the king of film. Mayer's office was
the size of a Zeppelin hangar and furnished in the style of Julius Caesar's
bedchamber. An overheated Shaw stood at the entrance with Gladstone in
hand, his tired whiskers drenched with sweat. The diminutive movie
mogul rose from behind his great desk, crossed forty yards of palatial carpet
to greet the dramatist, and grabbed Shaw's hand.

"I am so grateful you have come all this way to see us," said Mayer. "I
must say that I am a little embarrassed, since obviously you are here for art
and I am here for money."

Shaw was silent for a moment and replied, "Sir, I have not been
comfortable for one moment since I left home. I have traversed an ocean, a
continent, and your vast film factory. I have endured California sunshine.
I should have done none of these were I not here for money, while you are
here for art."

In the event, Shaw didn't reach terms with Mayer. Gabriel Pascal was
in the wings and the rest is movie history. The story sticks in my mind after
all these years because I have a hunch that Shaw's reply to Mayer had less
to do with art and money than with form and function. Mayer summoned
Shaw from England because he wanted the cleverest playwright alive;
Shaw was the master of form. On the other hand, Shaw came to Mayer to
reach the largest audience possible for his Fabian message; Mayer was the
master of function.

For an essay to work, form and function should be seamless. For an
essay to survive, form should outlast function. How dated the quarrels now
seem that sparked the great collections of yesteryear: not only Trilling's *The
Liberal Imagination*, but also Whitehead's *Science and the Modern World*,
Wilson's *To the Finland Station*, Barzun's *Teacher in America*! How timely the

minds of their authors remain! Margaret Fuller worried that the structure of an essay was dangerous because "it tempts to round the piece into a whole by filling up the gaps between the thoughts with - words - words - words." Her idol, Goethe, had it better when he hoped that his readers would come to understand that which was *not* written on the page: the life of its author. For whatever be the overt subject of an essay, it cannot avoid betraying some aspect of the writer's life. That's why a collection of essays tends to form an unintentional autobiography.

In this book of essays I have tried to fill "the gaps between the thoughts" not with words - words - words, but with memories. Memories that started when Mark Van Doren stood by his desk and stared out over our heads through the window at the frieze of great names carved into the facade of Butler Library.

1

DARWIN'S AUDUBON

In assessing Audubon, whose firm grip on the popular imagination has scarcely lessened since 1826, we must as historians of science seriously ask who would remember him if he had not been an artist of great imagination and flair. . . . The chances seem to be very good that had he not been an artist, he would be an unlikely candidate for a dictionary of scientific biography, if remembered to science at all.

ROBERT M. MENGEL, *Dictionary of Scientific Biography* (1970)

That pretty much describes how John James Audubon (1785–1851) is regarded as a scientist today. His name is missing from indexes of modern textbooks of ornithology, and our most eloquent natural scientist, Stephen Jay Gould, invokes Audubon only as a limner of "The Flamingo's Smile." Yet whatever honor is due Audubon as an artist of great flair and imagination—Winslow Homer regarded him the greatest of our water colorists—he was recognized by the most eminent of his contemporaries as their equal in natural science. But, alas, nowadays Audubon is honored more as a fan of the wild, a protoecologist, than as either the artist or the naturalist he considered himself. Leaving the art to others, I want to reassess whether Audubon should be "remembered to science at all" as his biographer suspects. By the standards of the day, he didn't do so badly. It was not for his art, but for his science that he was elected to all the learned societies; he became a Fellow of the Royal Society of London in 1830. Audubon shares that FRS with only one antebellum American, Benjamin Franklin—and with Newton and Boyle and the Darwins of the home team. His science as well as his art convinced the great Cuvier to introduce Audubon at the French Academy of Sciences in 1828, "The greatest monument yet erected by Art to Nature" said the Baron of the *Birds of America*. It seems appropriate that Audubon's tombstone in upper Manhattan was paid for and erected by the New York Academy of Sciences rather than its National Academy of Design.

But I've come across a real link between the work of Audubon and Charles Darwin, a link in that long chain of genetic reasoning that extends from Buffon to Lamarck to Audubon and to Darwin. The chain extends, of course, to the present when we have untangled DNA and cloned a small sheep in Scotland. A review of Darwin's correspondence, Audubon's journals, and a variety of primary sources yields direct evidence that Audubon's encyclopedic *The Birds of America* (London, 1827–38) contributed to Darwin's theory of natural selection. Moreover, not only Charles Darwin, but also Baron Cuvier, Geoffroy St. Hilaire, and others used the raw observations supplied by Audubon to support their notions of evolution. It is fair to say that Audubon's magnum opus, no less than the *Histoire Naturelle* (1770) of Buffon or *La Règne Animal* (1809–17) of Cuvier, was a necessary source for Darwin's *On the Origin of Species*. Evidence that *The Birds of America* was important for the invention of evolution is as close at hand as today's *Oxford English Dictionary* on CD-ROM. Inspired by a long look at Audubon's original watercolor of *The Magnificent Frigate Bird* at the New-York Historical Society, I plugged into the OED to learn about the frigate bird and related matters. Lo and behold, here sit Darwin and Audubon, embedded forever in silicon. Under "frigate" one finds "3. A large swift-flying raptorial bird (*Fregata aquila* or *Tachypetes aquilus*), found near land in the tropical and warmer temperate seas" followed by the citation: "1859 Darwin *Orig. Spec.* vi. (1878) 142: No one except Audubon has seen the frigate-bird . . . alight on the surface of the ocean."

What on earth was Darwin doing citing Audubon? "Found near land"—does that have any interest? It turns out that the Magnificent Frigate Bird is Plate 271 of the rare, hand-colored and hand-engraved double elephant folio of *The Birds of America*, while the text describing the bird is printed in Audubon's *Ornithological Biography*, separately published in five volumes (Edinburgh, 1831–49). The plates were published without text so as to avoid furnishing free copies to the public libraries of England, as copyright law demanded. Since only a score of the complete set of these books were available in England, Darwin must have had more than a cursory image in mind. He carefully studied both the engraving and the text, art as well as science. That passage of Darwin's from *On the Origin of*

Species which is cited in the OED tells us how important the frigate bird was for the theory of natural selection. I've got a hunch that the reason Darwin remembered Audubon's frigate bird "alight on the surface of the ocean" is that he needed that observation to refute an argument of Lamarck's.

Lamarck's "First Law," as announced in his *Philosophie Zoologique* (1809), was that use or disuse causes structures to enlarge or shrink. He illustrated this "use it or lose it" principle by showing how function might change structure in the digits of birds. He pointed out that those birds that need to find their daily prey in the water spread their digits and feet when they beat the water to move at its surface. Consequently, with time and over generations, the skin that unites these digits at their base would tend to shrink from this habitual movement of the digits in unison "and therefore those large membranes which unite the digits of the duck, the goose, et cetera come to assume their present-day forms." Lamarck's "Second Law" decreed that these changes were heritable, and "directed by forces which appear to us as a battle, but which is really nothing but necessity." Pangloss follows function, as it were.

But Darwin had seen the Magnificent Frigate Bird in Audubon's neo-classic splendor, swooping down on its prey, plunging into battle like the ship for which it is named. As befits a raptor, whose claws are aerodynamically trimmed for flight, the bird has retained only rudimentary webs on its feet. Audubon directed Havell, his master engraver, to transfer from water-color to the final colored print two closely detailed inserts of the dorsal and ventral aspects of the bird's feet with their residual webs. They are shown in red and yellow to document the age of the bird, and although the print and watercolor differ in the placement of these details, their persistence in the final plate indicates how important they were to Audubon. Indeed, of 435 plates in *The Birds of America*, only four still retain such anatomical inserts, and each is of the digits of a bird with webbed feet! Of those four, only the image of the frigate bird contains details that signal a specific stage of avian development.

Why was Audubon so interested in the feet of these birds? In his 1836 text he describes the swooping movements of the frigate bird in search of its prey: "They are most truly Marine Vultures. See him now! [He] approaches the water in the manner of the gulls!" That was the clue Darwin was to pursue, but Audubon was interested in another aspect as well: "In the third spring the upper parts of the head and neck of [the female] are a purer brownish black. . . . The feet, which at first were yellow have become a rich reddish orange. The bird is now capable of breeding." And in another passage of the *Ornithological Biography* he tells us that "I had been for years anxious to know what might be the use of the pectinate claws of birds." A modern scientist would describe the yellow-to-red transition as an example of ontogeny (how individuals develop). "Pectinate" claws are covered with scaly skin arranged in a basket-weave pattern which resembles two combs entwined. Audubon, in a small monograph published as an appendix to the *Ornithological Biography*, reasoned that such claws were used to scratch insects from the wings and might be expected in an upland bird that rarely alights on the water. But why does the frigate bird have both *pectinate claws*, useful for scrubbing insects inland, and the residua of *webbed feet*, useful for swimming in the water? Conclusion: the bird may represent a transition state between upland and seaborne birds. Audubon's rumination suggests that he became concerned—consciously or not—with phylogeny (how species develop). Darwin took up the challenge in *On the Origin of Species*

> What can be plainer than that the webbed feet of ducks and geese are formed for swimming? Yet there are upland geese with webbed feet which rarely go near the water; and no one except Audubon has seen the frigate-bird, which has all its four toes webbed, alight on the surface of the ocean. . . . The webbed feet of the upland goose may be said to have become almost rudimentary in function, though not in structure. In the frigate-bird, the deeply scooped membrane between the toes shows that structure has begun to change.

But what makes structure change? In the case of the frigate bird, its light, hollow bones and rudimentary digital webs should give it airborne advantage over its clumsier rivals. Its structure yields a survival advantage, the raptor's speed should serve to increase its number. In the very next paragraph, with Audubon's frigate bird in mind, Darwin enunciates his theory of natural selection:

> He who believes in the struggle for existence and in the principle of natural selection, will acknowledge that every organic being is constantly endeavouring to increase in numbers; and that if any one being varies ever so little, either in habits or structure, and thus gains an advantage over some other inhabitant of the same country, it will seize on the place of that inhabitant, however different that may be from its own place. Hence it will cause him no surprise that there should be geese and frigate-birds with webbed feet, living on the dry land and rarely alighting on the water.

Darwin goes on to quote Audubon at two other points in his text. In challenging Lamarck's first law, Darwin argues that the same species can accommodate to a wide variety of environments and that the same environment can support a wide variety of species: "I can only assert that instincts certainly do vary.... Audubon has given several remarkable cases of differences in the nests of the same species in the northern and southern parts of the United States." Finally, in refuting the notion that similar environments dictate similar habitants, Darwin calls on Audubon again. There is an easier explanation for why the same species can occupy widely distant, nonconnected bodies of fresh water: "I think it would be an inexplicable circumstance if water-birds did not transport the seeds of fresh-water plants to distant ponds and streams, situated at very distant points. ... I thought that the means of the dispersal [of large water-lily seeds] must remain inexplicable, but Audubon states that he found the seeds of the great southern water-lily in a heron's stomach."

I've checked out the index of Darwin's *On the Origin of Species*: among 1,440 references in that book, Lamarck, Geoffroy St. Hilaire, and even Erasmus Darwin only rate one mention apiece, but Audubon is quoted thrice, as often as Baron Cuvier. Indeed, Audubon is the only American or British ornithologist quoted other than Constantine Rafinesque (whose encounter with Audubon is one of the more amusing episodes in the *Ornithological Biography*). No citations of Audubon's predecessors or rival naturalists: no Charles Bonaparte, no William Swainson, no Ord, no Wilson, no Veillot, no Catesby, etc. Audubon alone appears essential to the Darwinian web-site.

Other citation counts confirm this notion. In his monumental *The Variation of Animals and Plants Under Domestication* (1868) Darwin quotes Audubon six times and refers twice to Audubon's and Bachman's less well known *Quadrupeds of North America* (1846). Remarkably, the index of Darwin's social treatise, *The Descent of Man and Selection in Relation to Sex* (1871), lists *The Birds of America* over two dozen times and Audubon and Bachman twice. Audubon beats out Buffon and Cuvier in each of these volumes.

But how did Darwin become familiar with the works of Audubon, and where did he gain access to Havell's masterly engravings of the watercolors for the double elephant folio? Although one can trace several direct and indirect connections between John James Audubon and Charles Darwin, their first encounter was in a year critical to both their careers. In 1826, the sixteen-year-old Darwin was a reluctant medical student at Edinburgh, more interested in long walks and curious rocks than in human anatomy. But family expectations were high: for two generations, medicine had been the family profession and natural science its passion. His grandfather, Dr. Erasmus Darwin (1731–1802), had been offered the post of physician to George III, but withdrew to Staffordshire to practice medicine, to form Birmingham's Lunar Society (with Wedgwood, Priestley, and Watt), to invent talking machines and pigment grinders, and to write treatises on Botany, on the Laws of Organic Life, and on Female Education. He was elected to the Royal Society in 1761. Charles Darwin's father, Robert, who was also a scientifically trained physician, literally became wedded to the industrial revolution; he married Josiah Wedgwood's daughter Susan. Robert Darwin's medical practice was vastly successful, while his microscopical interests earned him the second FRS in the family.

Young Charles was not a quick starter at Edinburgh. His belief that Dr. Robert Darwin would leave him property enough to subsist on "was sufficient to check any strenuous effort to learn medicine." But eventually he found a mentor in natural science in the person of Robert Lee Grant, a Lamarckian admirer of his grandfather's *Zoonomia*. He also attracted the attention of Robert Jameson, professor of natural history and president of Edinburgh's scientific society, the Wernerian. Jameson was the faculty advisor to the undergraduate Plini Society and persuaded Darwin to read his first two short papers before the Plinians. Darwin's "small discovery," that the so-called ova of the sea mat (a marine invertebrate) were actually ciliated larvae, was eventually picked up by Grant but Darwin "had not the satisfaction of seeing it in print" in his lifetime.

Audubon, age forty-one in 1826, was in the midst of a personal and professional metamorphosis from New World obscurity to European renown. His hopes for ornithological recognition and publication of *The Birds of America* had been dashed in the lyceums of New York and Philadelphia. With wife and sons left behind in the Louisiana bayous, he brought his portfolio of watercolors to England in search of patrons and a publisher. Owing to letters of introduction from his wife's family—the Bakewells had emigrated to America from England—he found a receptive audience among the gentry and naturalists of Liverpool and Manchester. Later that year he would go on to dine in triumph with the dons of the schools and the lords of the land. Edinburgh was one stage in that remarkable progression, and as chance would have it, one of Audubon's letters of introductions was to Darwin's Professor Jameson, who became Audubon's host at the university and sponsor of his membership in the Wernerian Society of Natural History. Jameson, the Wernerians, and other friends of Audubon arranged for public exhibition of the watercolors at the Royal Institution; these admission fees were Audubon's chief means of support. It was after a lecture by Audubon to the Wernerians on the turkey vulture or turkey buzzard (*Cathartes aura*) that Darwin wrote home dutifully: "it was there that I heard Audubon deliver some interesting discourses on American birds." Here is Audubon's description of these birds:

> Never spotted north of New Jersey. Mainly found near southern sea shore. Graceful in flight—wings spread beyond horizontal and their tips bent upward by the weight of their body. Gregarious but voracious eaters—they devour the young of many species and cannibalize their own.

> I have found birds of this species very old, with the upper parts of their mandibles, and the wrinkled skin around their eyes, so diseased as to render them scarcely able to feed amongst others .... I have represented the adult male in full plumage, along with a young bird, procured on the autumn of its first year. The average weight of a full grown bird is 6 1/2 lb, about 1 lb less than that of the carrion crow.

The language is sturdy, precise, and colloquial, it differs considerably from Audubon's Francophonic journal entries of the time. In fact, the *Ornithological Biography*, the text of which describes the plates of *The Birds of America*, was not composed by Audubon alone, but with the large editorial help of William MacGillivray. Some would indeed call MacGillivray the coauthor of Audubon's *Ornithological Biography*.

Charles Darwin was also in debt to this learned Scot. In his autobiography, Darwin recalls that "from attending Jameson's lectures, I became acquainted with the curator of the museum, Mr. MacGillivray, who afterwards published a large and excellent book on the birds of Scotland. I had much natural history talk with him and he was very kind." In Edinburgh, young Darwin had met both of the authors of *The Birds of America*.

Two years later, a still unsettled Darwin had quit medical studies to pursue divinity at Christ's College, Cambridge. Dallying between natural and moral philosophy he looked for a mentor to succeed Jameson and soon became both student and friend of Reverend John Stevens Henslow, professor of botany. Of the peripatetic botanist who changed his life forever, Darwin later wrote: "I have not as yet mentioned a circumstance which influenced my whole career more than any other. This was my friendship

J. J. Audubon, *Turkey Vulture (or Turkey Buzzard)*. New-York Historical Society, New York. © Collection of the New-York Historical Society.

with Professor Henslow ... I was called by some of the others 'the man who walks with Henslow.'" Henslow was a central figure in the Cambridge of his day; he established the Botanical Garden of the university, and together with the geologist Adam Sedgwick, FRS, founded the Cambridge Philosophical (Scientific) Society. Perhaps Henslow's most lasting contribution to science was this letter written on August 24, 1831, to the still floundering, twenty-two-year-old, Charles Darwin:

> *Henslow to Darwin*: August 24 1831 I have been asked by Peacock [professor of astronomy] to recommend him a Naturalist as companion to Captain Fitz-Roy, employed by government to survey the southern extremity of America. . . . Don't put any modest doubts or fears about your disqualifications, for I assure you I think you are the very man they are in search of; so conceive yourself to be tapped on the shoulder by your bum-bailiff and affectionate friend.

This Freudian invitation to share the voyage of the *Beagle* with Fitz-Roy was still three years in the offing when Audubon visited Cambridge in March of 1828. Audubon had arrived at Cambridge aiming to show his watercolors to the dons and to solicit subscriptions for the complete set of Havell's engravings for *The Birds of America*. A rara avis, himself, with his long locks and rustic garb, Audubon was courteously received by the Cantab naturalists and invited to dine at high table in Trinity College by William Whewell, FRS, professor of geology and future master of Trinity. Whewell, philosopher and mathematician, is best remembered today as the man who in 1840 was to coin the name "scientist" to replace the more archaic "philosopher," "natural philosopher," or "naturalist." He is also remembered by Darwin: "Dr. Whewell was one of the older and distinguished men who sometimes visited Henslow, and on several occasions I walked home with him at night."

After high table at Trinity, Audubon was taken to a private gathering. He wrote home to Lucy: "Professor Whewell took me to his rooms with some eight or ten others [including Henslow, Adam Sedgwick, et al.]. . . . Oh! My Lucy, that I also had received a university education! Whewell began asking me questions about the woods, the birds, the aborigines of America. We sat til late." He also wrote to his fellow ornithologist William Swainson: "How happy I should die when assured that science, by my feeble efforts, had advanced one step in its progress!" On the very next day, Audubon attended Sunday morning services at Great St. Mary's, where the sermon was delivered by none other than John Stevens Henslow. The paths of Charles Darwin, who walked with Henslow and Whewell, and of John James Audubon, who had dined with Whewell and Henslow, crossed again on the cobblestones of St. Mary's.

The trip to Cambridge was both a scientific and a financial success for Audubon. He had gained the attention of professors who were influential in English philosophical circles; they made possible his election to the Royal Society. His proponents were Professor J. G. Children of the British Museum and Lord Stanley of the Zoological Society, each of whom had ties to the dons gathered in Whewell's chamber, and each of whom subscribed to *The Birds of America*. Henslow himself persuaded the University Library to acquire one complete set of the double elephant folio and the Philosophical Society to purchase another; a substantial commitment of Cambridge funds. Audubon was to remain in correspondence with Henslow for several years, successfully urging Henslow to acquire the "best and most durable" bindings for the engravings. And there in Cambridge they remain today, finely bound in morocco, as Waldemar Fries assures us in his bibliography of *The Double Elephant Folio*. In return, Audubon remembered his friends in natural science: ornithologists owe to Audubon not only "MacGillivray's Finch" but also "Henslow's Bunting." One might say that John Stevens Henslow not only helped to launch *The Voyage of the Beagle* but also to hatch *The Birds of America*.

There is yet another link of Audubon to Darwin. There was a direct connection between Darwin's grandfather and Lucy Bakewell Audubon. Before the Bakewells emigrated to America, Dr. Erasmus Darwin had been their family physician. The Bakewells lived in Matlock, a hamlet between Manchester and Derby, and Dr. Darwin was buried twelve miles away towards Derby. In 1826, shortly before that lecture on the turkey vulture which Charles Darwin attended, Audubon had visited Erasmus Darwin's tomb. Audubon knew and admired Dr. Darwin's *Zoonomia*, but he was not simply paying homage to a fellow naturalist, the connection was familial. He had promised Lucy that "I will return to Manchester . . . visit thy native Matlock; gaze on the tomb of the friend of thy youth, Darwin." Audubon paid that visit and decorated his journal entry with a sketch of the bridge over the Derwent: "To Matlock, where Dr. Darwin nursed thee on his knee."

Erasmus Darwin, FRS, was friend and doctor as well to most of the Lunar society who formed the matrix of the industrial revolution into which Charles Darwin was born: Joseph Priestley (of oxygen), James Watt (of the steam engine), Josiah Wedgwood (of pottery), Matthew Boulton (of steam-

J. Wright, *An Experiment with the Air Pump*. The National Gallery, London.

powered smelting), and Joseph Wright of Derby (of *An Experiment with the Air Pump*). Wright's neoclassical painting illustrates not only the physical and medical side of the "Lunaticks"—science gives one power over life and death—but also refers to its everyday application. The air pump greatly amused Dr. Darwin and Priestley because when the apparatus is used in reverse, it can be made to carbonate water. Nowadays, Priestley is known to bottlers as "the father of the soft drinks industry." Erasmus Darwin, who was an inveterate fan of useful devices and treatments, was curious whether bottled water might have medicinal value. In 1794, his friend, the industrialist Matthew Boulton, replied with a reference that rings a bell today. Boulton replied:

> J. Schweppe [a jeweler in Geneva] prepares mineral waters of three sorts. No.1 is for common drinking with your dinner. No.2 is for nephritick patients and No.3 contains the most alkali given only in the most violent cases.

Wright's neoclassical painting, which illustrates Priestley's air pump, alludes to a less amusing link between Audubon and the Lunaticks. It seems that Priestley came up in many of Audubon's tall tales. Priestley, Lucy Bakewell and family, and Audubon, all settled in Pennsylvania in the wake of the French Revolution and its various reverberations in France, Haiti, and the West of England. Priestley was honored more in absentia in the parlors of the Manchester gentry than when he had championed the French Revolution. And when Audubon showed his English hosts letters of introduction from the Bakewells, he did not hesitate to claim personal acquaintance with the great chemist. It was a fib on a par with his false claim to have studied with Jacques-Louis David; neither has served him well over the years in censorious eyes.

The familial links between Darwin and Audubon tend to center around Dr. Darwin's *Zoonomia*. Written in the heroic couplets of the day, it is a Voltairean encyclopedia of creation. The good doctor applied the reductionism of Newton to living things. He argued that creation was brought about not by God, but by the working of physical forces; first HEAT, then REPULSION, then ATTRACTION, and

> *Last, as fine goads the gluten-threads excite*
> *Cords grapple cords, and webs with webs unite,*
> *And quick CONTRACTION with ethereal flame*
> *Lights into life the fibre-woven frame,*

In branching cones the living web expands
Lymphatic ducts, and convoluted glands;
Aortal tubes propel the nascent blood,
And lengthening veins absorb the refluent flood;
Leaves, lungs and gills the vital ether breathe
On earth's green surface, or in the waves beneath.

Perhaps in consequence of the cultural sea-change that followed the end of the Enlightenment, Charles Darwin's *Origin of Species* shows no great respect for freethinking Dr. Darwin. But in 1826, the year young Darwin heard Audubon in Edinburgh, Charles had not yet become the complete Victorian. Charles Darwin recalled walking with Dr. Grant of the Wernerians:

> [Grant] burst forth in high admiration of Lamarck and his views of evolution, I listened in silent astonishment, and as far as I can judge, *without any effect on my mind.* I had previously read the *Zoonomia* of my grandfather, in which similar views are maintained, *but without producing any effects on me*.... At this time I admired greatly the *Zoonomia*; but on reading it after an interval of ten or fifteen years, I was much disappointed; the proportion of speculation being so large to the facts given [my italics].

It might strike a modern reader that *Zoonomia* did in fact produce an effect on the mind of the grandson. Charles Darwin acknowledged that his grandfather had anticipated the doctrine of evolution, but that the good doctor had fallen into the Lamarckian error of assuming that "use and disuse" created separate species. Nowadays, some of Dr. Darwin's speculations seem remarkably—shall we say?—Darwinian. Erasmus Darwin certainly anticipated the doctrine of the survival of the fittest; in *Zoonomia* (I:395–6) he proposed that "the strongest and most active animal should propagate the species, which should thence be improved." Nevertheless, his grandson found room to cite Dr. Darwin only once in his text, and as a footnote to Lamarck, at that:

> [Lamarck] attributed the gradual change of species . . . much to use and disuse, that is, to the effect of habit. To this latter agency he seems to attribute all the beautiful adaptations of nature;—such as the long neck of the giraffe for browsing on the branches of trees. [Footnote] It is curious how largely my grandfather, Dr. Erasmus Darwin, anticipated the views and erroneous grounds of opinion of Lamarck in his *Zoonomia* vol I pp 500–501, published in 1794.

In this context we may recall that the full title of Darwin's great opus is *On the Origin of Species by Means of Natural Selection or the Preservation of Favoured Races in the Struggle for Life*. In our anachronistic days, references to Favoured Races smack of social Darwinism, as celebrated in Haeckel's *Darwinismus* [German] which kindled much of the mischief of our own century. But for Darwin and the Victorians, Nature was the book in which the laws of empire were written. What was "necessity" for Lamarck in the eighteenth century became the order of battle to the Victorians of the nineteenth century: a glorious war of nature.

Audubon at Edinburgh chose as his example of nature red in tooth and claw the turkey vulture, set to "devour the young of many species and cannibalize their own." The image is likely to have stirred the heart of an uncertain medical student pained by the sight of blood. In that critical year of 1826, in an operating theater in Edinburgh, young Darwin "saw two very bad operations, one on a child, but I rushed away before they were completed." He never returned to the clinic. Perhaps it is no accident that Darwin used another of Audubon's raptors, the Magnificent Frigate Bird—that "truly Marine Vulture"—to refute Lamarck and his own grandfather. Lamarck believed that evolution obeyed forces "which appear to us as a battle but which are really nothing other than necessity"; Darwin taught us that battle is the necessary force. The difference between Lamarck and Darwin is the difference between the eighteenth and nineteenth centuries.

Years later, Charles Darwin recalled walking with Grant to the Wernerians; they spoke of Lamarck and of the *Zoonomia*: "it is probable that the hearing rather early in life of such views maintained and praised may have favored my upholding them under a different form in my *Origin of Species*." That different form, based on close observation of the natural world, contained in what Darwin considered the proper proportion of speculation to fact. In the reciprocal mode, I'm persuaded from the many links between Darwin and Audubon—but especially from the impact made on Darwin by the Magnificent Frigate Bird and the turkey vulture—that the theory of natural selection is based, in some small measure, on the facts of nature as collected by Audubon.

Darwin drew on Audubon for clues not only to natural, but also to sexual selection. We often forget that the full title of Darwin's other great book is *The Descent of Man and Selection in Relation to Sex*. Darwin depended greatly in this opus on Audubon's vivid accounts of avian courting rituals to argue that the mating game had genetic consequences. Darwin deduced that the "selected" winner of the game passed heritable elements to his progeny by virtue of innate fitness. Those transmissible genetic elements were contained in the "germ plasm" as August Weismann was to call it in 1885. Indeed, Darwin wrote the foreword to the English translation of Weismann's work on sexual selection and the germ plasm. And after Weismann's "continuity of the germ plasm" had been properly traced to DNA by Avery, MacLeod, and McCarty (1944) the way was clear to the double helix and—for better or worse—to Dolly.

In the last paragraph of *Origin of Species*, Charles Darwin asks us to imagine an entangled bank:

> clothed with many plants of many kinds, with birds singing on the bushes, with various insects flitting about, and with worms crawling through the damp earth . . . . Thus, from the war of nature, from famine and death, the most exalted object which we are capable of conceiving, namely, the production of the higher animals, directly follows. There is grandeur in this view of life, with its several powers, having been originally breathed into a few forms or into one; and that, whilst this planet has gone cycling on according to the fixed law of gravity, from so simple a beginning endless forms most beautiful and most wonderful have been, and are being, evolved.

Perhaps the best illustrator of that entangled bank has been John James Audubon.

Speaking in the voice of Audubon, the American poet Pamela Alexander replies:

> *I am a collection of landscapes,*
> *a Gazetteer. Of griefs & hopes*
> *& consummations. I must be old now, although*
> *I think so only in the company of mirrors . . .*
> *[I] wandered & drew Quadrupeds*
> *& birds my passion*
> *& as they do not pay*
> *took faces of strangers in chalk, painted*
> *street signs & a small weasel*
> *to be graved for bank notes. Collected*
> *for my America its wildness, made distant grass & near Grebe*
> *to shine alike with detail,*
> *clear as water . . .*

The churr & chuck of blackbirds is my work song.
Stiff pointed wings of hawks produce
a whistling sound, & blunt Duck wings whir.
I have heard my shot strike the Frigate bird in flight
& then the slapping of its wings in the disorder
of mortal fall.

2

INFLAMMATION AS
CULTURAL HISTORY

Recherche scientifique: la seule forme de poésie qui soit rétribuée par
l'État.
[Scientific research: the only sort of poetry that is state-supported.]

JEAN ROSTAND, *Inquiétudes d'un biologiste*

Redness and swelling with heat and pain—*rubor et tumor cum calore et dolore*—have been recognized as the four cardinal signs of inflammation since the writings of Cornelius Celsus in the first century of the common era (30 B.C. to 38 A.D.). Celsus—who is sometimes mistaken for Celsius of thermometry—was describing the typical reaction of flesh to microbes. Although all sorts of injuries to humans and beasts will elicit inflammation, it seems clear that our extensive arsenal of host defenses has not been stocked by evolution against such recent threats as ionizing radiation, crack cocaine, bombs plastique, or Katyusha rockets. No, the drab Darwinism of biology suggests that whereas inflammation may help the individual cope with cuts and bruises, most of the redness and swelling with heat and pain is there to make sure that the species is not wiped out by epidemics. It seems to me that the more we learn about inflammation, the simpler its message becomes: Our cells and humors defend the self against invisible armies of the other. We call our losses "infection" and our victories "immunity."

As in other branches of science, the language we use to describe the battle of inflammation derives as much from our cultural heritage—or the temper of our times—as from the facts of nature. Thus, for example, Ernest Renan has properly traced the clarity of Claude Bernard's prose to the Jansenist "*maitres de Port-Royal.*" Indeed, we cannot read Bernard's description of the action of curare, "*curare détruit le mouvement mais reste sans action sur le sentiment; il dissèque en quelque sorte le système nerveux moteur"* ["curare halts movement but not sensation; dissecting by this means the motor nervous system"], without being reminded of the spare rhetoric of Pascal and—ultimately—the *Sic et Non* of Abelard.

The warfare between humans and bacteria was not appreciated as an order of battle until the microbe hunters of the nineteenth century recognized the opposing forces and the territory in dispute. By 1908, when the Nobel prize was given to Paul Ehrlich for his work on humoral immunity (antibodies) and to Élie Metchnikoff for his work on cellular immunity (phagocytosis), it was clear that the body uses these two strategies in concert to identify and destroy invaders. Metchnikoff (1905) announced that discovery in the discourse of Darwinian survival:

> When the aggressor in this struggle is much smaller than its adversary the result is that the former introduces itself into the body of the latter and destroys it by means of infection. . . .But infection also has its counter. The attacked organism defends itself against the little aggressor. It protects itself by interposing a resistant membrane, or it uses all the means at its disposal to destroy the invader.

The tendency to couch descriptions of inflammation in terms of nineteenth-century battle has proved to be irresistible. Indeed, the assumptions of the microbe hunters were based not only on models of Darwinian zoology but also on the military legends of empire. In 1941, the distinguished pathologist Joseph McFarland summed up the field of inflammation from Rudolf Virchow to Valy Menkin:

> To many, the situation here encountered resembles a battlefield on which the leukocytes meet the invading bacteria and contest their further increase and invasion until they triumph, and, the infection overcome, the inflammation subsides.

The microbe hunters drew their images of battle from romantic accounts of skirmishes waged more often than not by splendidly equipped British or French troops against primitively armed "lesser breeds." Appropriately, Anatole France remarked that *"le principe fondamentale de toute guerre colonial est que l'Européen soit supérieur aux peuple qu'il combat: sans quoi la guerre n'est pas coloniale, cela saute aux yeux"* ["the fundamental principle of a colonial war is that the Europeans are superior to the people they are fighting, otherwise it is not a colonial war, that's obvious"]. Those colonial wars were directed by distant generals who viewed the destruction of foreigners through the lenses of binoculars. One catches the whiff of inflammation in John Bowle's account of General Kitchener's revenge (1896) for the uprising in the Sudan that killed the noble Gordon:

> The campaign culminated in the battle of Omdurman . . . when some 60,000 of the Khalifa's horde flung themselves with superb courage against Kitchener's line, to be mowed down by machine guns, rifles, and artillery. . . . By 11:30 A.M. Kitchener handled his binoculars to an aide de camp, remarked that the enemy had a "thorough dusting" and ordered the advance on Omdurman and Khartoum. The victory had cost 48 killed, including 3 British officers and 25 other ranks and 434 wounded, as against over 11,000 "dervish" dead and about 16,000 wounded and prisoner. Gordon had been more than avenged; the British and Egyptian flags again floated over Khartoum, and when, at the service of thanksgiving the troops sang "Abide with me," Gordon's favorite hymn, even Kitchener was seen to weep.

Substitute "endothelium" for "line," replace "machine guns, rifles, and artillery" with "macrophages, leukocytes, and fixatives (antibodies)," and we can appreciate the language that was used in the nineteenth century to describe how higher organisms deal with the dervishes of microbe. Metchnikoff, Ehrlich, Cohnheim, and Adami peered through the lenses of microscopes to watch battalions of white cells avenge the revolt of bacilli. Caught in the Victorian structures of hierarchy, Metchnikoff, a gentle scholar, taught the lesson of Darwinian phylogeny from the evidence of phagocytosis:

> The diapedesis of the white corpuscles, their migration through the vessel wall . . . is one of the principal means of defense possessed by an animal. As soon as the infective agents have penetrated into the body, a whole army of white corpuscles proceeds towards the menaced spot, there entering into a struggle with the micro-organisms. The leukocytes, having arrived at the spot where the intruders are found, seize them after the manner of the Amoeba and within their bodies subject them to intracellular digestion.

Metchnikoff was not the only champion of white corpuscles; we owe their classification as basophils, eosinophils, and neutrophils to Ehrlich's doctoral thesis on aniline dyes. Blue, red, and white colors depended on the way in which positive or negative charges lined up in the granules of the phagocytes; aniline dyes marked the combatants as clearly as red or blue tunics identified cavalry units in the Franco-Prussian War of 1870. Indeed, aniline dyes were the response of German synthetic chemistry to French control of the colonies that yielded natural dyestuff. Ehrlich's studies with dyes, which introduced the language of colors and fixation, helped Germany to become the arsenal of chemotherapy and had the happy side effect of launching immunochemistry.

Ehrlich elaborated the dictum that *Corpora non agunt nisi fixata* ("Bodies that do not attach do not act") and this principle became the basis not only for the first treatment of syphilis—Salvarsan—but also for modern ligand–receptor theory. It might be argued that Ehrlich's doctrine also applies directly to the means whereby circulating cells of the blood (white cells and platelets) stick to vessel walls and to each other in the course of inflammation; we call those processes "heterotypic" and "homotypic cell adhesion," respectively. In recognition of our newfound mastery of the molecules that regulate these movements (integrins, selectins, and their ligands), we might nowadays rephrase Ehrlich's dictum as *Cellulae non agunt nisi fixata* ["Cells that do not attach do not act"].

Central to the humoral theories of Ehrlich and the cellular ones of Metchnikoff was the late Victorian conviction that the body would not injure itself wittingly. As the nation-states of Europe placed their faith in the social doctrine of Darwinian survival, the microbe hunters were equally sure that the body had a *horror autotoxicus*—an incapacity to turn defensive weapons into tools of self-destruction. Time and events have overturned that nineteenth-century conviction. The bad news of the twentieth century—from Flanders' fields to the fall of Paris, from the siege of Madrid to the Berlin Wall, from the battle of Algiers to the mess in Yugoslavia—has added anxiety to the orderly prose of biology. Nowadays when we speak of inflammation, we speak of unplanned mischief, as Lewis Thomas did in his first essay on inflammation (1971):

> I suspect that the host is caught up in mistaken, inappropriate, and unquestionably self-destructive mechanisms by the very multiplicity of defenses available to him, defenses which do not seem to have been designed to operate in net coordination with each other. The end result is not defense; it is an agitated committee-directed, harum-scarum effort to make war, with results that are remarkably like those sometimes observed in human affairs when war-making institutions pretend to be engaged in defense. If, to push the analogy, there were no limit to the number of people who could set off for northern Minnesota at the season of the great fly-over of geese and no limit on the type and power of the weapons to be used by each, we would undoubtedly observe . . .

with M-16's, howitzers, SAM missiles, lasers, and perhaps tactical nuclear rockets considerably more destruction of people than geese, of host than invader.

The change in our opinion that inflammation is benign can be traced to the first Great War of this century. It may not be a complete coincidence that the terms "allergy," "anaphylaxis," and "serum sickness" began to crowd the clinical literature while the medical profession was prepping the national armies of Europe for slaughter in the trenches of 1914. The immunologists of Europe were prepared to believe that cells could respond to inflammation as readily by revolt as by enlistment. And the grinding offensives of the First World War succeeded not only in trimming the rhetoric of battle of operatic glamour but also in striking the set of nationalism. Class struggle began to dissolve the ties of patriotism as the effects of that war of attrition spread over Europe. It soon became apparent that World War I had dismantled the ceremonies of class upon which the Victorian nation-state had relied. The captains and the kings had killed each other off, leaving corporals to rule a diminished world. Ford Maddox Ford recognized the changing of the guard in *Parade's End*:

> He was devising the ceremonial for the disbanding of a Kitchener battalion . . . Well the end of the show was to be as follows: The adjutant would stand the battalion at ease, the band would play Land of Hope and Glory, and then the adjutant would say: "there will be no more parades. . . . For there won't . . . No more Hope, no more Glory, no more parades for you and me anymore. Not for the country . . . nor for the world, I dare say . . . None . . . Gone . . . No . . . more . . . parades!"

Carrying forward the analogy we have drawn between the discourse of politics and the discourse of biology, we might say that science in the 1920s and 1930s learned about histamine, serotonin, enzymes, and peptides—widely diffused mediators of inflammation—as the nations learned about other mediators of inflammation: fascism, Stalinism, and the Nazis. Biologists were taught by Sir Henry Dale and Sir Thomas Lewis that human cells can synthesize and release chemicals as disabling to the organism as any toxin released by microbes. That unpleasant lesson coincided with one of the few points of doctrine agreed upon by parties of both right and left: that the body politic carries the seeds of its own destruction. No longer were human differences drawn by national borders alone; to the separations of class were added the divisions of ideology. Again, it may be no accident that we learned about shock lung and the crush syndrome (now attributed, at least in part, to activation of anaphylatoxins like C5a in the circulation) from Loyalist surgeons in the Spanish Civil War. That rehearsal for the world war against fascism reached a stalemate at the medical school in Madrid in November of 1936—as described by Hugh Thomas:

> Hours of artillery and aerial bombardment, in which neither side gave way, were succeeded by hand-to-hand battles for single rooms or floors of buildings. In the Clinical Hospital, the Thaelmann Battalion [of the Republican International Brigades] placed bombs in the lifts to be sent up to explode in the faces of the Moroccans [of the Franco forces] on the next floor. And, in the next building, the Moroccans suffered losses by eating inoculated animals kept for experimental purposes.

Nevertheless, until the civil warfare in Madrid, it had been clear on which side of the battle line each soldier stood. It was in Madrid that the fascists boasted of a "fifth column"—secret right-wing partisans who would join with the four regular columns of Franco troops to conquer the republic. The terms "right," "left," "communist," and "fascist" had a generally accepted meaning of sorts. But when the Second World War came and the Germans occupied most of Europe, even those boundaries became blurred. Distinctions among classes, ideologies, nationalities, religions, and genders became most confused, and especially so in the colonies of France occupied by the Germans. A model of this confused state was André Gide. Gide, the gay, ex-communist littérateur, was a friend of de Gaulle and Malraux on the one hand and of cultivated Germans on the other; his ambivalent journal records the advance of Eisenhower's troops on Tunis, then occupied by the Germans:

> Several trustworthy farmers confirm the lamentable, absurd retreat of the American Force before the semblance of German opposition. The sudden appearance of a handful of resolute men forced the withdrawal of those who, very superior in numbers and equipment, would have had only to continue their advance to become masters of their objective . . . Tunis. Germans everywhere. Well turned-out, in becoming uniforms, young, vigorous, strapping, jolly clean-shaven, with pink cheeks. . . . Can there be a more wretched humanity than the one I see here? One wonders what God could ever possibly come forth from these sordid creatures, bent over toward the most immediate satisfactions, tattered, dusty object and forsaken by the future. Walking among them in the heart of the Arab town, I looked in vain for a likable face on which to pin some hope: Jews, Moslems, southern Italians, Sicilians,

or Maltese, accumulated scum as if it were thrown up along the current
of clear waters.

From a similar matrix of "accumulated scum" and "jolly" Germans,
Hollywood distilled the movie *Casablanca*. That tough tearjerker set the
measure of language for my generation. Redness and swelling with heat
and pain seemed to be the cardinal emotions of *Casablanca*, and since Bogey
was the last hero we could trust, we were sure that all its passions were
worthwhile. But in the dreary half-century since the film, we have learned
to distrust our leaders, our heroes, and our wars. We have gotten very good
only at rounding up the usual suspects.

In science, we have isolated and defined all sorts of new substances but
have been disappointed to find that our own cells and fluids collaborated
with hostile invaders to provoke inflammation. Sometimes these Quislings
and Lavals turn on us in the absence of any enemy. In exposing the treasons
of lymphocytes, we have come to view them as a cast of Levantine characters. In our film-derived discourse, Peter Lorre launders the money for the
contrasuppressor T cell, and Sydney Greenstreet cashes in the loot at the
bursa of Fabricius. Controlled by a network of codes, could we call the signs
of genetics anything else? The alliances of inflammatory cells shift, and their
affinities wane; wounds heal but scars remain.

The new synthesis makes only superficial sense. Like the *Marseillaise* in
the film, the humors of immunity flood our spirit. Our psyches are lifted or
depressed by the products of inflammation; we have learned that the battle
cries of injury (interleukins, interferons, tumor necrosis factor) make us
febrile, debilitated, sleepy, and cross even as they arouse battalions of
lymphocytes. Related molecules activate the troops of defense: phagocytes,
which wear receptors on their sleeves as if they were the armbands of local
resistance. Paul Henreid leads the forces of that interior.

But in this postbacterial, postmodern world, the phagocytes fail, white
cells do damage, and our antibodies form complexes (the very coin of
neuroses) and plug our kidneys. In the course of these confused responses,
much harm is done and resolution is not invariably achieved. Every few
years we add a new felon to the list of usual suspects: chemokines, substance P, anti-idiotypes, leukotrienes, lipoxins, and the appropriately
named free radicals. Now that the cold war is over, we speak neither the
language of war nor that of peace; we can no longer decide if our tissues
are entirely inflamed or partially immune. Our self-destruction is brought
about by molecules that dozens (or more) kinds of invaders have learned

to put at their service (such as interleukin 1 and tumor necrosis factor). In the discourse of politics and science, we have replaced defined suspects with uncertain suspicions. Church and state dismiss modern science by accepting its procedure, but not its ethic. The high sentiment of *Casablanca* has yielded to the flat affect of the *policier*, Malraux and romance have been replaced by Simenon and the *dossier*. Matters are no better in the world of politics: Jacques Monod was correct to complain that

> *Armées de tous les pouvoirs, jouissant de toutes le richesses qu'elles doivent à la science, nos sociétés tentent encore de vivre et d'enseigner des systèmes des valeurs déjà ruinés, à la racine, par cette science même.*
>
> [Armed with all the power, enjoying all the riches that science has given, our societies remain content to live with and to teach value systems that, at root, have been ruined by science itself.]

Indeed, the *systèmes des valeurs* of our time are not exactly in ruins, but accommodate to the culture of Disneyland. We have turned from the rational branches of knowledge taught by the philosophes to become pluralistic, multicultural, and multiracial. We have become not members of a sex but part of a gender, instead. Our Western culture has become nontraditional in every sense: baubles of other lands, as well as rhythms of other folk, compete for attention with the high art of our time. We have learned the charms of Benin masks, Shaker quilts, Dayak chants, Zuni pots, samurai gear, and the Andean flute, Indian dances, and the Grateful Dead. Malraux's museum without walls has become a souk without borders. So rich is the polyglot feast that we despair of taking it all in: we may become so filled by hors d'oeuvres that we leave before the main course.

Science has also burst its limits. We are taught political behavior by students of ants and are taught the rules of conception by lawyers in court. We define human death by a brainwave machine and define the onset of life by judicial appeal. We now blame each disease in the textbook on our grandfather's genes but pardon the treason of microbes. For every cure we effect, a new virus crawls out of the marrow to haunt us. The networks of memory and immunity intersect, and the plans for their wiring resemble the Northeast American power grid—or vice versa. We have, or have not, mastered superconduction and tabletop fusion, but—as in the recent Benveniste affair—we won't know if the data are fudged until the editor of *Nature* sends a magician to check out the notebooks.

The effect of our new molecular biology is—like the Norfolk of Noël Coward—very flat. We have learned that the hierarchy of genes is controlled by runs of DNA (called homeoboxes) which do not set us apart from fruit flies. All the signals of chromosomes seem to be able to work "on the other side this time." Every codon has an anticodon: every strand of DNA can be read for sense in one direction and perhaps even greater sense in the other. Recently we were assured by the *Journal of Immunology* that "each of us has a receptor for all antigens. . . . Individuals, then are a composite and a reflection of all universal shapes represented in the antigen repertoire to which they respond."

In other words, we recognize the invader—call him antigen or Ishmael—and recognize him because we have his template in our genes, but—like the Scarlet Pimpernel—suddenly on the other side. We are told that we have seen the enemy and he is coded, in language a fruit fly can understand, in our genetic strands. We get no good guide to the shift in our cultures from experts. Cultural rock-stars like Foucault and Derrida have written dodgy scenarios to follow: the discourse of art and the gaze (*le regard*) of science are contingent or transient, each reader decodes a new text, and a phrase is a sign with ten meanings. Poetry is born in misreading, and science conceived in fraud. The dons of DNA offer no greater consolation. Having discovered that the gears of our genes mesh quite well in reverse, they have slipped new pieces of genes into cells to control them. These nucleic acid constructs are called—you might have guessed this from Roland Barthes—"antisense" strands of RNA or DNA. Their instructions are read by blobs of proteins with "zinc fingers" or "leucine zippers." The structure of language is analogous to the structures of nucleic acids, capable of transcription, translation, and betrayal—on either side, in any direction.

But stop! I'm not quite ready for that kind of diminished script just yet. No, I'm ready to cheer again for that last shot of Humphrey Bogart and Claude Rains walking off to Brazzaville to join the forces of Charles de Gaulle. I want to hear the swelling music that plays at what could be the start of a beautiful friendship. I want to believe that all that redness and swelling with heat and pain is up to some good. I want to set sail with Baudelaire on a voyage that leads to the new, and if death be the captain, so be it:

> O Mort, vieux capitaine, il est temps! levons l'ancre.
> Ce pays nous ennuie, ô Mort! Appareillons!
> Si le ciel et la mer sont noirs comme de l'encre,
> Nos coeurs que tu connais sont remplis de rayons!

Verse-nous ton poison pour qui'il nous réconforte!
Nous voulons, tant ce feu nous brûle le cerveau,
Plonger au fond du gouffre, Enfer ou Ciel, qu'importe?
Au fond de l'Inconnu pour trouver du nouveau!

[O Death, old captain, it's time! raise the anchor.
This country bores us, O Death! Let's set sail!
Though sea and sky be dark as ink
Our hearts you know are awash in light!
Pour us your poison for that will refresh us
We will, as long as that fire scorches our brain,
Dive deep into the abyss, heaven or hell, what matters?
At the end of the unknown to discover the new!]

3

GERTRUDE STEIN AND THE CTENOPHORE

The photograph shows students in the summer dress of a century ago collecting specimens at low tide from a harbor near Woods Hole in Massachusetts. The harbor is Quisset and its waters today remain rich in marine life; its heights are still dominated by a Yankee cottage called "Petrel's Rest"; the house is still surrounded on four sides by a veranda and fronted by a flagpole. The young people are on a collecting trip for the course in Invertebrate Zoology at the Marine Biology Laboratory; its students still collect specimens from the inlets of Buzzard Bay. The photo was taken on July 31, 1897: The young woman in the middle is Gertrude Stein. She has turned, smiling to her brother Leo, who holds up a specimen jar: "Look what I've found!" he gestures to the photographer. Many in the group are also smiling, it is the height of summer; they are young and have disembarked at a marine Cythera where lush creatures drift on tides warmed by the Gulf Stream. Leo Stein has snared a ctenophore, a solitary, free-swimming member of the hydromedusae: a stunning trophy from the sea.

At Woods Hole, Gertrude Stein was twenty-three; she had just finished Radcliffe and was to enter the Johns Hopkins Medical School in the fall. That summer she had enrolled in the embryology course of the MBL and often accompanied Leo and the invertebrate zoologists on collecting trips. She was on her way to becoming almost, but not quite, a doctor. Not every student at the MBL goes on to a career in medicine or science, some may not even learn much science, but few will forget those moments at the beach or at the bench when the cry goes up: "Look what I've found!" W. C. Curtis, who is shown in the photograph in white cap and knee-boots just to the right of Stein, remembered in the *Falmouth Enterprise*: "For us that summer she was just a big fat girl waddling around the laboratory and hoisting

Gertrude Stein and Her Brother Finding Ctenophores on Quisset Beach, 1897. Marine Biological Laboratory Archives, Woods Hole, MA.

herself in and out of the row-boats on collecting trips." He might have added that at the time she was still in thrall to her fast-talking brother: some might read the photo as a record of a very intense bond. Stein has written that as children she and Leo had tramped alone in the woods of northern California; her adoring gaze in that photo from Woods Hole suggests an American future for them both—from the redwood forests to the Gulf Stream waters, as it were.

But the photo suggests motifs other than those of a family romance; the image alone is a stunning icon of natural science. The anonymous photographer has snapped a tableau of figures in a landscape so arranged that all the compositional lines—from hillside on the upper left to beach grass at the lower right—converge on Leo who holds high the collecting jar with his discovery. The creature from the ocean has been plunked into a glass pot to become a specimen for science. The composition has elements of Joseph Wright's neoclassic *An Experiment with the Air Pump* in that the figures are so disposed as to lead us to the apex of a visual pyramid at which life has been caught in a jug. The photo leads our eye, via the sightlines of Gertrude and another young woman, straight to Leo who looks to the lens with his prize held high. Looking at the photo a century later, we know that the scales of discovery tipped toward the sister and therefore we tend to read the picture as an action shot of the artist as a young woman. There, on the beach at Quisset, at the dawn of a century whose art she will help form, she looks forever happy by a summer sea.

The picture also evokes memories of darker days in the middle of the twentieth century. The solitary, shimmering ctenophore, trapped in a jar, recalls aspects of Gertrude Stein other than her art. Frederick Dupee characterized her as "at once female and male, Jew and non-Jew, American *pur sang* and European peasant, artist and public figure, and so on," while Catherine Stimpson asks her Rutgers students more bluntly: how did two Jewish lesbians survive the Nazis in France? As we've learned from a recent volume of her letters, Stein and her companion, Alice B. Toklas, were by no means trapped by the war, but voluntarily sat out the German occupation in a French alpine resort. In line with her modernist infatuation with fascist politics in general, and her support of Franco's rebellion in particular, she spent the war years translating a collection of Maréchal Pétain's speeches for an American audience, praising the Nazi puppet as the "George Washington" of France. Be that as it may, between the beach at Quisset and the 1940s at Culoz, she changed forever the way we read the English language.

No other picture of the young Stein shows her quite as perky as on that day at Quisset. A group photo of her embryology course, taken later in the summer, shows her preoccupied and unsmiling. Indeed, nothing about science or medicine seems to have given her much joy after her tussle with ctenophores. After indifferent attention to laboratory and clinic, she failed to graduate from the Johns Hopkins Medical School with her class of 1901. Although she had completed the bulk of her work, she seemed to have floundered over obstetrics—she who had gotten all that marine embryology right. She tells us that it was in the course of her obstetrical work that she became "aware of the Negroes" in Baltimore clinics serviced by Johns Hopkins; from that experience emerged the story of Melanctha in *Three Lives*. The mulatto abandons a humdrum doctor in favor of more louche companionship. To hear her tell it, Gertrude Stein and Johns Hopkins separated as if by mutual consent. In *The Autobiography of Alice B. Toklas* she relates:

> The Professor who had flunked her asked her to come to see him. She did. He said, of course, Miss Stein all you have to do is to take a summer course here and in the fall naturally you will take your degree. But not at all, said Gertrude Stein, you have no idea how grateful I am to you. I have so much inertia and so little initiative that very possibly if you had not kept me from taking my degree I would have, well, not taken to the practice of medicine, but at any rate to pathological psychology and you don't know how little I like pathological psychology, and how *all medicine bores me*. The professor was completely taken aback and that was the end of the Medical education of Gertrude Stein. [my italics]

Well, not exactly. Stein did not fail pathological psychology, but obstetrics. Moreover, her life at the Johns Hopkins of William Osler—who defined female medical students as "the third sex"—may have been many things, but it could not have been boring. The minutes of the Medical Faculty Advisory Board record that "the Dean presented the marks of the Graduating Class and on motion of Dr. Osler it was voted that Miss Gertrude Stein be not recommended for the degree of Doctor of Medicine." Whether from boredom with medicine or deeper battles of the self, no joy shows in her picture with the class of 1901 at Hopkins. She stands glumly half-hidden in the back row behind other women and the shorter, swarthier of the men. She left for Europe, and after aimless wanderings with her brother, settled in Paris to find her own vocation. Gertrude Stein soon outdistanced Leo who had stopped dabbling in biology; she discovered new art and new

Class Photo of the Marine Biological Laboratory Embryology Course, 1897. Marine Biological Laboratory Archives, Woods Hole, MA.

friends on the banks of the Seine. By the time Picasso painted her in 1906, her Iberian portrait showed the confident young writer who was crafting *Three Lives* in the course of becoming—in the words of Carl Van Vechten—midwife to the twentieth century.

Those who look to details of biography for explanations of literary or artistic styles can usually extract as much material as needed to convince. Stein was born on the north side of Pittsburgh in 1874 to a prosperous German-Jewish clothing merchant; the family business was dissolved in the year of her birth. Her father took the Stein family abroad for years of continental wandering and early education, before returning to Oakland where a new family cable business was established. Although orphaned in adolescence, Stein was assured of an independent income for life; many of her American connections, first in San Francisco and then in Baltimore, were to the kin of wealthy textile merchants.

Stein's haute-bourgeois, international background, which she shared with Henry James, has prompted Dupee and others to call attention to the mandarin elements her work shares with the Master. But it was Henry's brother, William (M.D. Harvard, 1869), who seems to have played a greater role in Gertrude Stein's career. The most provocative explanation of how her unique style developed—its ontogeny as it were—was offered by B. F. Skinner, writing in *The Atlantic* of 1935. In an article entitled "Has Gertrude Stein a Secret?" the Harvard professor of psychology traced Stein's technique to her undergraduate research work on automatic writing with William James, Skinner's predecessor as professor of experimental psychology at Harvard. The case is persuasive, even if Skinner's aesthetic verdict on Stein's later work is a tad harsh. But another influence, that of Jacques Loeb, her instructor at Woods Hole, has not escaped notice. In the formative years of her compositional style, Gertrude Stein owned that she was an old-fashioned mechanist, a reductionist of the school of Jacques Loeb. Her revolution of words owed as much to Loeb's *The Mechanistic Conception of Life* as to her study of "Normal Motor Automatism," which she had written for William James, or to the great *Demoiselles d'Avignon* of Picasso, the first versions of which showed a medical student as spectator.

When Stein trod the beach at Quisset, Jacques Loeb (1859–1924) was the leader of a new, mechanistic school of American biology which tried to explain the phenomena of biology by the equations of physics. In 1897, Loeb was teaching the physiology course at Woods Hole and the implications of his biophysical approach were the talk of the MBL. He had demonstrated that the chemical nature of salts in the environment of a cell controlled its irritability, movement, and reproduction in predictable ways. He was on his way to creating life in a dish, to forming fatherless sea urchins by chemical means. Parthenogenesis was announced two years later, but his work on tropisms and salt solutions had already paved the way for what Loeb the called the fundamental task of physiology: "to determine whether or not we shall be able to produce living matter artificially."

Loeb and his school of mechanists believed that they were the legitimate heirs of the philosophes and Loeb's book *The Organism as a Whole* (1916) is dedicated to Denis Diderot in the words of John Morley's tribute to the philosophe: "He was one of those simple, disinterested, and intellectually sterling workers to whom their personality is as nothing in the presence of the vast subjects that engage the thoughts of their lives." The mechanist's credo, with its belief that it is possible to frame disinterested thoughts unshaped by "personality," has since been repeatedly dismissed as shallow reductionism, but modern science has devoted itself to fulfilling Loeb's prophecy of "producing mutations by physico-chemical means and nuclear material which acts as a ferment for its own synthesis and thus reproduces itself." Indeed, it can be argued that much of modern art—from cubism, to constructivism, to abstraction, to minimalism, to pop art—has aimed to eliminate "personality" from the "vast subject" of form itself.

But Loeb claimed more territory for science than that of the body. He extended its empire to the mind. Replying to William James' request for his views of brain function, Loeb responded: "Whatever appear to us as innervations, sensations, psychic phenomena, as they are called, I seek to conceive through reducing them—in the sense of modern physics—to the molecular or atomic structure of the protoplasm, which acts in a way that is similar to (for example) the molecular structure of the parts of an optically active crystal" (1888).

Loeb addressed this reductionist proposal to the one man perhaps least likely to be persuaded. William James had argued eloquently the opposite view before an audience of Unitarian ministers at Princeton in a lecture entitled "Theism and the Reflex Arc." In a ringing defense of theism against the reductionists of the reflex arc, he told the liberal clergy that:

> Certain of our positivists keep chiming to us, that, amid the wreck of every other god and idol, one divinity still stands upright,—that his name is Scientific Truth. . . . But they are deluded. They have simply chosen from the entire set of propensities at their command those that

were certain to construct, out of the materials given, the leanest, lowest, aridest result,—namely the bare molecular world, —and they have sacrificed all the rest. . . . The scientific conception of the world as an army of molecules gratifies this appetite [for parsimony] after its fashion most exquisitely.

Stein got a whiff of both sides in the course of her research career. She had heard the siren song of vital forces from the voice of gentle James; she worked with one of Loeb's closest colleagues, Franklin Pierce Mall, not only in the embryology course at Woods Hole, but also at Johns Hopkins. But by the time she published *Three Lives*, her critics got it just right: she was on the side of the mechanists. Wyndham Lewis—another modernist fan of fascism—argued in his review that Stein's book put demotic speech into the "metre of an obsessing time" and although "undoubtably intended as an epic contribution to the present mass-democracy" gave "to the life it patronizes the mechanical bias of its creator."

The careers of Stein and Loeb ran somewhat in parallel. Orphaned in youth, both were brought up by well-off relatives as secular Jews, Stein in Oakland and Loeb in the Rhineland. Both emigrated as young adults, both spoke their adopted languages with awkward accents. Early in their careers, both performed experiments on brain function that caught the attention of William James; Loeb had written to James in search of a faculty position in the New World. When Stein entered Hopkins, she worked on mechanical models of spinal tracts with Frank Mall, Loeb's colleague from Woods Hole and the University of Chicago; both Stein and Loeb engaged the far from casual interest of B. F. Skinner. Skinner was persuaded by Loeb's argument in *The Organism as a Whole* that mechanist principles could be applied to the study and control of behavior. Skinner was also captivated by *Arrowsmith*, in which Loeb—as Martin Arrowsmith's mentor, Dr. Gottlieb—is depicted as a secular saint of science. Skinner was a true believer in Loeb/Gottlieb's *Mechanistic Conception of Life* and it surprised no one that when he turned his attention to Gertrude Stein, he deciphered her work in the language of nerves.

Skinner was the first to draw attention to Gertrude Stein's undergraduate work on automatic writing. He pointed out the origins of her verbal experiments by unearthing the paper on "Normal Motor Automatism" published in the *Psychological Review* of September 1896 by Leon M. Solomons and Gertrude Stein from the Harvard Psychological Laboratory (Wm. James, professor). Solomons and Stein reported on experiments designed to test whether a second personality—as displayed in cases of hysteria—could be called forth deliberately from normal subjects. The two authors undertook to see how far they could "split" their personality by eliciting automatic writing under a variety of test conditions. They concluded that hysteria is a "*disease* of the *attention*" (their italics), basing their argument on the finding that when distracted or inattentive, normal subjects show the abnormal motor behavior of hysterics. It may be no coincidence that Solomons and Stein performed laboratory work on hysteria in the very year that Sigmund Freud and Josef Breuer published their clinical *Studies in Hysteria*, 1895. The subject was much in the air on both sides of the Atlantic, not least because—as William J. McGrath has suggested—the study of hysteria offered science an opportunity to strike at the foundations of religion. Explain away divine madness by the reflex arc and you explain away divinity. James as a theist was persuaded that experimental psychology would validate all the varieties of religious experience. His students, on the other hand, came to a behavioral conclusion that favored the reflex arc: wrote Solomons and Stein, "An hysterical anesthesia or paralysis is simply an inability to attend to sensations from this part."

Skinner showed little interest in Solomons and Stein's discussion of hysteria, being more concerned with tracing Stein's literary style to her experiences of automatic writing. Using themselves as test subjects, Solomons and Stein were able to show that with a little practice they could regularly produce automatic writing as they took dictation while reading another text: "The word is written or half-written before the subject knows anything about it, or perhaps he never knows anything about it. For overcoming this habit of attention we found constant repetition of one word of great value." After they had trained themselves by this sort of cognitive drill, and after sessions with Ouija boards to call up their alter egos, automatic writing became easy. Stein found it convenient to read what her arm wrote, but following it three or four words behind her pencil. In this fashion, "a phrase would seem to get into the head and keep repeating itself at every opportunity, and hang over from day to day even. The stuff written was grammatical, and the words and phrases fitted together all right, but there was not much connected thought."

Skinner—a traditionalist with respect to the arts—argued that these experiments explained why there appeared to be two Gertrude Steins. The first Stein was accessible, and had written such serious work as *Three Lives* and *The Autobiography of Alice B. Toklas*; the other Stein was dense, and wrote stuff that was grammatical, with words and phrases fitted together all right but without connected thought. The second Stein had written *Tender Buttons*, her portraits, and *The Making of Americans. Four Saints in Three Acts* was yet to come! Skinner gave mild positive reinforcement to the first Stein, but strong negative reinforcement to the second, chiding her that "the mere generation of the effects of repetition and surprise is not in itself a literary achievement." Skinner complained that the second Stein gives no clue as to the personal history or cultural background of the author and dismissed her most adventurous book with a phrase from their common master, William James: "*Tender Buttons* is the stream of consciousness of a woman without a past." It is, of course, next to *Q.E.D.*, her most homoerotic work; no past indeed!

On the surface, Skinner seems to have scored a point. It is easy to pick up resonances between the samples of automatic writing that Solomons and Stein present and the matrix of Stein's later work. Thus the first two passages below, examples of automatic writing from the 1896 article, *sound* like the second passages from later Stein; a closer look will show all the difference:

> 1. "This long time when he did this best time, and he could thus have been bound, and in this long time, when he could be this to first use of this long time . . ."
>
> Solomons and Stein

> 2. "When he could not be the longest and thus to be, and thus to be, the strongest . . ."
>
> Solomons and Stein

> 3. "One does not like to feel different and if one does not like to feel different then one hopes that things will not look different. It is alright for them to seem different but not to be different."
> Stein, *Meditations on Being About to Visit My Native Land* (1934)

> 4. "What a day is today that is what a day it was day before yesterday, what a day!"
> Stein, *Broadcast on the Liberation* (1944)

Skinner's theory of the two Steins permitted him to "dismiss one part of Gertrude Stein's writing as a probably ill-advised experiment and to enjoy the other and very great part without puzzlement." The irony seemed to have been lost on our leading reductionist that he was reducing Stein's new style to its leanest, lowest, aridest origin: the knack of automatic writing she had acquired in the course of her undergraduate experiments.

Other interpretations might occur to those who believe that behavior—not to speak of literature—might be described in more complex, dynamic ways. William James wrote Stein to acknowledge receipt of *Three Sisters*: "I promise you that it shall be read *some* time! You see what a swine I am to have pearls cast before him! As a rule reading fiction is as hard to me as trying to hit a target by hurling feathers at it. I need *resistance*, to cerebrate!" What a challenge to fiction by the brother of Henry James! Hurling feathers, indeed! But, of course, Gertrude Stein had not simply written yet one more work of fiction, she had fabricated a new language. Stein's language turned the quiet elisions of homoerotic love into a seamless web of modern prose, no wonder she encountered enough resistance for a gaggle of Jameses.

Her incantation to the Liberation: "What a day is today that is what a day it was day before yesterday, what a day!" is not only pure Stein but can also be read as an utterance of disguise. In either case, it is couched in the language of our century: short repetitive sequences. Short repetitive sequences run through modern literature from Morgenstern to Vonnegut, from Beckett to Pinter; they charge the beat of modern music from rock to Philip Glass. Short repetitive sequences also constitute the language of our genes, when we talk DNA or RNA we speak pure Stein. We would not be surprised to hear a molecular biologist explaining a stretch of DNA in the one-letter code of nucleic acids: What a TAA is ATAA that is ATA what a TAA it was, what a TATAA, what a TAA, what a TATAA! Gertrude Stein worked in the manner Loeb attributed to Diderot: she wrote disinterested sentences the sound of which no false note of personality was permitted to disturb. Her champions praise her as the last daughter of the eighteenth century and the herald of the twentieth (Brinnin). Like the cubists, she had broken the common plane of thought to make compositions from its basic verbal elements. As Loeb reduced the structure of living things to the "bare molecular world," so Stein reduced language to its bare molecular level where phonemes throb to their own rhythm. Stein's new language expanded the technology of prose as cubism expanded that of painting.

Gertrude Stein tells us that she had learned from William James that "science is continuously busy with the complete description of something, with ultimately the complete description of everything" (*Lectures in America*, 1935). But words, sounds, *things* were not descriptions: "A daffodil is different from a description, a jonquil is different from a description. A narrative is different from a description. . . . A narrative is at present not necessary" (*How to Write*, 1931). Neither modern writing nor modern art were going to *describe* things, that was the job of science. Nor should the moderns write narrative, the last century had smothered words with stories. No, the task of writers in the twentieth century was to free words and images from the baggage of sentiment. Free from myth, meaning, and station, each jonquil or daffodil, each magpie or pigeon, could stand fresh on the page. A rose is a rose is a rose—and with that line, said Stein, a rose was red for the first time in English poetry for one hundred years. She had reinvented the rose and was free to create an army of roses at will, each red for the first time.

But Stein's story, which began with Loeb and the ctenophore at Quisset, ends on a note that reverts to a murkier mood of William James. James—who had found the Civil War too rich for his blood—had assured the Unitarians that

> if the religion of exclusive scientificism should ever succeed in suffocating all other appetites out of a nation's mind . . . that nation, that race will just as surely go to ruin, and fall a prey to their more richly constituted neighbors, as the beasts of the field, on the whole, have fallen a prey to man. . . . I myself have little fear for our Anglo-Saxon race. Its moral, aesthetic, and practical wants form too dense a stubble to be mown down by any scientific Occam's razor that has yet been forged. The knights of the razor will never form among us more than a sect.

Gertrude Stein and Alice Toklas sat out the grimmer stages of Hitler's War in a villa equipped with servants in the town of Culoz, 50 kilometers over the border from Geneva. They appear to have been protected from the roundup of Jews and foreigners by the prominent French fascist Bernard Faÿ, who served Pétain and his puppet government as director of the Bibliothèque Nationale. Faÿ also edited the monthly *Les Documents maçonniques* which was devoted to rooting out freemasonry, or belief in "exclusive scientifism," as James would have it. From July 1940 until June 1944, Faÿ was also an editor of the only journal financed by the Germans, the anti-Jewish *La Gerbe*. Tried and convicted as a war criminal, he nevertheless obtained Stein and Toklas' letters of support during the immediate postwar period; he had taken care of their art collection during their stay in the Alps. Evidence has recently come forth that Stein was no unwilling "victim" of the Nazis; she undertook to introduce a collection of Pétain's speeches with this outrageous comparison:

we did not understand defeat enough to sympathize with the French people and with their Maréchal Pétain, who like George Washington, and he is very like George Washington because he too is first in war first in peace and first in the hearts of his countrymen, who like George Washington has given them courage in their darkest moment held them together through their times of desperation and has always told them the truth and in telling them the truth has made them realize that the truth would set them free.

Perhaps the worst untruth she perpetuated was her bland assurance that the French were all darlings throughout the war:

Speaking of all this there is this about a Jewish woman, a Parisienne, well known in the Paris world. She and her family took refuge in Chambery [a stone's throw from Culoz where Stein and Toklas shopped each Tuesday] when the persecutions against the Jews began in Paris. And then later, when there was no southern zone, all the Jews were supposed to have the fact put on their carte d'identité and their food card, she went to the prefecture to do so and the official whom she saw looked at her severely Madame he said, have you any proof with you that you are a Jewess, why no she said, well he said if you have no actual proof that you are a Jewess, why do you come and bother me, why she said I beg your pardon, no he said I am not interested unless you can prove you are a Jewess, good day he said and she left. It was she who told the story. Most of the French officials were like that really like that.

Well, not exactly. Within a few blocks of M. Faÿ's office at the Bibliothèque, French officials rounded up hundreds of French "Jewesses" and their children to pack them off in cattle cars. "*N'Oubliez pas!*" the stone tablet reads today at the Marché des Blanc-Manteaux. Many of Stein's coterie turned their backs on these events, while not a few were Pétain sympathizers or worse. It is no accident that in *Axel's Castle*, Edmund Wilson ranks Stein with Yeats, Pound, Eliot, and Valéry. Like Stein, the doyens of prewar modernism linked a taste for avant-garde literature with one for *arrière-garde* politics. Again like Stein, many of the French modernists—from Cocteau to Gide, Peguy to Claudel, Lartigue to Maillot; indeed, Van Dongen to Matisse—behaved during the Nazi occupation as if the struggles of our time were simply an unpleasant interruption to their life in art. It may also not have been an accident that when Gertrude Stein was asked about the bomb that ended the war against Japan, she replied: "They asked me what I thought of the atomic bomb. I said I had not been able to take any interest in it."

Perhaps such attitudes as "boredom" with medicine or "disinterest" in nuclear weapons are the high price some artists pay to gladden their muse. Or perhaps the strict homophobia of William Osler and the late Victorians has much to answer for. Whatever. In the light of her undoubted genius, the tribute Stein paid to Picasso can be read as her own ambiguous lesson of self-love and hate:

> a creator is contemporary, he understands what is contemporary when the contemporaries do not yet know it, but he is contemporary and as the twentieth century is a century which sees the earth as no one has ever seen it, the earth has a splendor that it never had, and as everything destroys itself in the twentieth century and nothing continues, so then the twentieth century has a splendor which is its own and of this century, [it has] that strange quality of an earth that one has never seen and of things destroyed as they have never been destroyed as they have never been destroyed. [*Picasso*, 1938]

4

PUERPERAL PRIORITY

*I do know that others had cried out with all their might against the
terrible evil [of puerperal sepsis] before I did and I gave them full credit
for it. But I think I shrieked my warning louder and longer than any of
them and I am pleased to remember that I took my ground on the
existing evidence before the little army of microbes was marched up to
support my position.*

Letter of Oliver Wendell Holmes to Wm. Osler (1894)

As we have learned from the recent histories of AIDS, of Lyme disease, and
of the Ebola virus, two medical factions tend to compete for honors in the
war against infectious disease. One favors epidemiology and statistics
while the other depends on laboratory science to find and kill the little army
of microbes. Our past victories over infection have required help from both
contenders—sanitarian and microbe hunter we might call them—but the
honors we assign vary with time and place.

Who conquered childbed fever? The French credit Pasteur for finding
the microbe that caused it; England honors Lister for antisepsis; Europeans
acclaim Semmelweis for the etiology and prophylaxis of puerperal fever;
the United States gives priority to Oliver Wendell Holmes. In Scotland yet
another is honored, and justly so. In 1905 the University of Aberdeen
painted on the wall of one of its classrooms: "The infectious nature of
puerperal fever was first demonstrated by Dr. Alexr. Gordon Aberdeen,
1795." There is, of course, glory enough for all, sanitarian and microbe
hunter alike; before their efforts, epidemics of puerperal fever killed up to
a third of young mothers. Nevertheless, at the dawn of the twentieth
century an eminent obstetrician lamented in the *British Medical Journal* that

2,000 women were lost each year in England and Wales "from a cause which is almost, if not entirely, preventable, and that puerperal fever continues to prevail, as though Pasteur and Lister had never lived." Why did the epidemics of childbed fever persist for most of the nineteenth century, why did so many women die after Pasteur and Lister?

In 1995 UNICEF estimated that 100,000 women would die from puerperal sepsis somewhere in the world—as though Pasteur and Lister, not to speak of Fleming, had never lived. The history of childbed fever may shed light on why those "almost, if not entirely, preventable" diseases caused by streptococci, spirochetes, or retroviruses persist today.

GORDON AND HOLMES

Two hundred years ago, Dr. Alexander Gordon of Aberdeen knew that puerperal fever was a contagious disease of which physicians or midwives were the chief vectors: "I had evident proofs that every person who had been with a patient in the puerperal fever became charged with an atmosphere of infection which was communicated to every pregnant woman who happened to come within its sphere." He found it disagreeable "to mention that I myself was the means of carrying the infection to a great number of women." Based on the literature and his personal observation, he concluded that "these facts fully prove that the cause of the puerperal fever . . . was a specific contagion, or infection, altogether unconnected with a noxious condition of the atmosphere." But his pamphlet raised no greater stir in the profession than had the claims of others; Gordon's argument was ignored by his colleagues who believed in the zymotic theory of infection: "a noxious condition of the atmosphere."

By 1843 a more powerful voice was raised. Dr. Oliver Wendell Holmes had returned to Boston from medical studies in Paris, a city which before the Civil War attracted the best of American medical students. Holmes was drawn to Paris not only by the abundant clinical material and unlimited opportunities for dissection that the city offered, but also by a new Gallic spirit that promised to transform medicine into a science based on number, measure, and observation. Chief proponent of the new method at the École de Médecine was Professor Pierre C.-A. Louis, whose work on what we now call "patient outcomes" discredited blood-letting as a treatment for infection. Louis founded the Society of Clinical Observation to which Holmes was admitted and the principles of which the young American brought home. The motto of Louis and his society became the principle of quantitative medicine in Boston:

Formez toujours de idées nettes.
Fuyez toujours les à peu pres.
[Formulate your ideas precisely.
Always avoid the approximate.]

The thirty-two-year-old Holmes was a known quantity when he brought clinical observation home to Boston. Son of a Harvard divine, as an undergraduate he had saved the US frigate *The Constitution* from the scrapheap. (His poem "Old Ironsides"—*Aye, tear her tattered ensign down, etc.*—was widely circulated at the time and until a generation ago remained alive in American medical schools as its parody: *Aye, tear her tattered enzyme down, etc.*) His career as a medical scholar was launched at the Boston Society for Medical Improvement before which he read "The Contagiousness of Puerperal Fever." He argued persuasively that "the disease known as puerperal fever is so far contagious as to be frequently carried from patient to patient by physicians and nurses." He began by reviewing the medical literature, giving full marks to Gordon of Aberdeen: "[Gordon's] most terrible evidence is given in these words: 'I ARRIVED,' he says, 'AT THAT CERTAINTY IN THE MATTER, THAT I COULD VENTURE TO FORETELL WHAT WOMEN WOULD BE AFFECTED WITH THE DISEASE UPON HEARING BY WHAT MIDWIFE THEY WERE TO BE DELIVERED, OR BY WHAT NURSE THEY WERE TO BE ATTENDED, DURING THEIR LYING-IN; AND ALMOST IN EVERY INSTANCE MY PREDICTION WAS VERIFIED '" (Holmes' caps).

Holmes next reviewed Registrar's statistics, estimating that the mortality in England of puerperal sepsis varied at the time from 3 to 5/1,000, a figure, Holmes calculated, in keeping with its incidence in Boston. He contrasted this low, sporadic level with certain epidemics arising from one focus: a Dr. Ruter had reported to the College of Physicians of Philadelphia in 1842 that he had personal experience of nearly seventy of these "horrible cases" all within the past twelve months. Holmes commented "in view of the general incidence of the disease being less than five in a thousand, it does appear a singular coincidence, that one man should have ten, twenty, thirty or seventy cases of this rare disease following his or her footsteps with the keenness of a beagle." Another Philadelphia physician, Dr. Warrington, reported that a few days after assisting at autopsy in a case of puerperal peritonitis in which he ladled out the contents of the abdominal cavity with his hands, he delivered three women in rapid succession; all were afflicted with puerperal fever. Soon afterwards he saw two other patients, both on the same day, with the same disease: of these five patients, two died. Holmes also recounted "the well-known story of Dr. Campbell of Edinburgh," who, in October 1821, "assisted at a *postmortem* examination in a case of puerperal fever, carrying the pelvic viscera in his pocket to the classroom. The same evening he attended a woman in labour, and he was called to a second the following morning. Both these patients died, as well as three others whom he delivered within the next few weeks."

Holmes anticipated the deductive methods of Sherlock Holmes, who was named after him by Conan Doyle. He contacted doctors in whose practices puerperal sepsis had appeared in clusters and was able to figure out how the contagion spread. One man wrote Holmes about a "disastrous period in my practice," when between July 1 and August 13, 1835, he attended at the delivery of fourteen women; eight of these developed puerperal fever and two died. He admitted that he had not changed his outer coat until August 6, the date of confinement of the last victim! Holmes studied two other such focal outbreaks and rightfully concluded that "the occurrence of three or more cases, in the practice of one individual, no others existing in the neighborhood, and no other sufficient cause being alleged for the coincidence, is *prima facie* evidence that he is the vehicle of contagion." Holmes had fulfilled the task he set himself: "The practical point to be illustrated is the following: *The disease known as Puerperal Fever is so far contagious as to be frequently carried from patient to patient by physicians and nurses*" [Holmes' italics]. The countermeasures he urged were those of the sanitarian revolution: plenty of soap and water, destruction of contaminated clothing, linens, or blankets; doctors or nurses should not deliver babies for several months after an outbreak.

Holmes' warning caused few ripples outside New England; the small medical journal in which he published went under within the year and Holmes himself moved on to other pursuits. Public medical belief had been formed by the doyens of Philadelphia who denied that physicians might carry the pestilence. Professor Hodge's popular textbook urged doctors to "divest your minds of the overpowering dread that you can ever become ... the minister of evil; that you can ever convey, in any possible manner, a horrible virus, so destructive in its effects, and so mysterious in its operations as that attributed to puerperal fever." The influential Professor

Charles Meigs refused to consider that a clustering of cases in one practice might be evidence of its spread by contagion: "I prefer to attribute them to accident, or Providence, of which I can form a conception, rather than to a contagion of which I cannot form any clear idea, at least as to this particular malady."

The general inattention to his earlier warning led Holmes in 1855 to republish and elaborate his work on puerperal fever. By then, Holmes had become a stronger force in the profession; he was not only the Parkman Professor of Anatomy at the Harvard Medical School, but also dean of its medical faculty. It was Holmes who gave the name "anesthesia" to Morton's discovery (1846). He had also become a respected poet and lyceum lecturer; his monthly essays in *The Atlantic*—to be collected in *The Autocrat of the Breakfast Table*—had already made him a household name among literate folk. In his new pamphlet, he directed harsher words at his main opponents and marshalled statistics against them: "three deaths in one thousand births [may be] assumed as the average from puerperal fever; no epidemic to be at the time prevailing. I have had the chances calculated by a competent person, that a given practitioner, A., shall have sixteen fatal cases in a month [data from Philadelphia records], it follows . . . there would not be one chance in a million million million millions that one such series would be noted." Holmes' peroration was memorable: "The teachings of the two Professors in the great schools of Philadelphia [Hodge and Meigs] are sure to be listened to [and] I am too much in earnest for either humility or vanity, but I do entreat those who hold the keys of life and death to listen to me also for this once. I ask no personal favor; but I beg to be heard in behalf of the women whose lives are at stake, until some stronger voice shall plead for them."

This time the case was overwhelming, as was its acceptance by the American medical community. Dr. Elisha Harris in New York took Holmes' side in his review of Meigs' treatise on childbed fever; Meigs had taken pains to deny the contagious etiology of puerperal fever and advised bloodletting as its treatment. Wrote Harris: "The poetic Harvard professor has furnished an absolute demonstration by facts, of the affirmative, while Meigs has defended the negative with the poetic arguments of fanciful hypothesis." Puerperal fever had long plagued New York's oldest and largest public hospital, Bellevue. There the disease often reached epidemic proportions and Dr. Fordyce Barker found that there were times when "a pulse of 120 and a flushed cheek were looked for as a matter of course on the morning after confinement, and the normal results were luxuries to the attendant physician."

After Holmes' work became widely know, the situation changed. At the New York Academy of Medicine in 1857, Dr. Alonzo Clarke described how puerperal fever had broken out at Bellevue Hospital that April. Following Holmes' precepts, the infected women were removed, the place cleaned, new furniture installed, and a different set of doctors and nurses placed in charge. The following month Clarke delivered twenty-three women, not one of whom contracted the fever. Dr. Barker, in commenting upon Clarke's paper, not only agreed with him that the physician was often the medium of infection but also pointed out that Dr. Holmes had demonstrated this "many years earlier with an array of facts which must . . . be convincing to every unprejudiced mind," and that the work of Ignaz Semmelweis in Vienna fully bore out Dr. Holmes' observations. The sanitarian revolution had planted its flag in New York!

Philadelphia alone remained unconvinced, or at least until 1889, when—we are told by the home page of its university—"whether out of concern for modesty, infection, or other reasons, a committee reportedly was organized in Philadelphia to set fire to the newly established university maternity hospital soon after its completion." But in most of the United States, the verdict of William Osler was sustained: "so far as this country is concerned, the credit of insisting upon the great practical truth of the contagiousness of puerperal fever belongs to Dr. Holmes."

SEMMELWEIS AND LISTER

Dr. Barker, like Holmes and other Americans, knew the work of Semmelweis from accounts by English students who had been in Vienna. But although Holmes quoted Semmelweis in 1855, the opposite was not the case, neither in the Hungarian's preliminary abstracts of 1848 nor in the full publication of 1861, probably because Semmelweis, like other Central Europeans, regarded the New World as terra incognita with respect to medical science. A pity, since as Dr. Barker knew, while Semmelweis and Holmes were an ocean apart, their findings were complementary. Holmes had deduced contagion by his retrospective study of private practice and he advised relative *asepsis* as the remedy; Semmelweis studied contagion prospectively in a charity ward by testing *antisepsis* as the remedy. Yet, while their sanitarian conclusions as to the cause of childbed fever were similar, the lives and circumstances of Oliver Wendell Holmes and Ignaz Phillip Semmelweis were antipodal: One was a Boston Brahmin, the other a grocer's son at birth; one was an honored sage, the other a mental patient at death; one wrote compellingly, the other was prolix to tedium. Holmes, unlike Semmelweis, lived to see his work accepted by his colleagues the world over, who—in Auden's rhyme for Paul Claudel—pardoned him for writing well. Not so Semmelweis; a medical bibliographer has argued that if the Hungarian "had written like Oliver Wendell Holmes, his *Aetiologie* would have conquered Europe in twelve months."

Unique conditions prevailed at the Algemeiner Krankenhaus in Vienna where the young Semmelweis served as house physician in obstetrics. Before 1840, the two sections of that department at the hospital had approximately equal mortality rates from puerperal fever; but beginning in 1840, the male medical students were all assigned to the first section, while the female midwifery students were assigned to the second section. Deaths in the first section exceeded those in the second by four- or fivefold. In the first section mortality sometimes approached twenty percent, that in the second section rarely exceeded one percent.

Semmelweis correctly deduced that more women died in the first section because only the medical students attended autopsies; Semmelweis himself had assisted Rokitansky at autopsies of infected patients. He drew a connection between the lesions of puerperal sepsis and those found in a colleague of his, Jakob Kolletschka, who died from pyemia (septicemia) contracted at the autopsy table. He concluded that there was something contagious in the "fetid material" of the autopsy room and that thorough cleaning of the hands might prevent the toxin from spreading: "In the examination of pregnant or delivering maternity patients, the hands, contaminated with cadaverous particles, are brought into contact with the genitals of these individuals, creating the possibility of resorption. With resorption, the cadaverous particles are introduced into the vascular system of the patient. In this way, maternity patients contract the same disease that was found in Kolletschka. Only God knows the number of patients who went prematurely to their graves because of me."

To stop the spread of "cadaverous particles," Semmelweis made the medical students wash their hands with chlorinated limewater; beginning in May of 1847 the mortality rates in the first division dropped from 18.3 to 1.3 percent, and so effective were his methods that between March and August of 1848 no woman died in childbirth in his division. In the spring of 1848, however, the liberal revolution which swept Europe evoked a nationalist, xenophobic backlash. Probably because advancement in the profession became restricted to ethnic Austrians or Germans, the Hungarian Semmelweis was dismissed from his post at the clinic. Nevertheless, he made two brief and unheralded presentations to the Viennese Medical Society on puerperal sepsis and on that basis applied for a teaching post in obstetrics. It was no surprise that he and the university failed to come to terms.

Returning to Hungary, Semmelweis repeated his successful attack on childbed fever at the St. Rochus hospital in Pest where he worked for the next six years. Although in nearby Prague the mortality rates of puerperal sepsis remained over ten percent, Semmelweis reduced the rate at St. Rochus to less than one percent. Alas, Vienna remained as resistant to Semmelweis as Philadelphia had been to Holmes. Championing zymotic notions, the *Wiener Medizinische Wochenschrift* ridiculed the provinces and urged an end to chlorine hand washes. Predictably, mortality rates from puerperal sepsis rose again into the teens.

In 1860, Semmelweis published his principal work, *Die Ätiologie, der Begriff und die Prophylaxis des Kindbettfiebers*, and sent it around to the prominent medical societies of the day. The general European response to his summa was adverse or dismissive; and some historians have traced the subsequent clouding of his mind to that rejection. A review of *Die Ätiologie* leaves little doubt that Semmelweis had become mentally troubled a few years earlier. His treatise, either in German or in English, is somewhat of a mess, presenting raw data in clotted prose. After publication of his magnum opus, Semmelweis became even more dispirited, and drifted into an ill-defined mental syndrome. The circumstances of his death in an Austrian mental institution (1865) are also unclear. His encyclopedist claims that he died of septicemia after an operation: "He died of the same disease against which he had struggled all his professional life." Others believe that he died from an infected wound received as he was taken in custody. And although his torch was briefly raised again by Mayrhofer—who defined childbed fever as a "fermentation disease" cause by living vibrions—by the end of 1865, Semmelweis and his work were in eclipse. Public belief had prevailed over public health.

Semmelweis returned to posthumous favor along with the microbe hunters, not because he had found how the germ was spread and by whom, but because he was a pioneer of antisepsis. When the Regius Professorship

of Surgery at Glasgow University fell vacant in 1865, Joseph Lister—son of
J. J. Lister FRS, a pioneer microscopist—was elected from among seven
applicants. The Glaswegians, who had recruited Lister to a renovated Royal
Infirmary, hoped that operative sepsis, which sometimes claimed over half
the surgical patients between 1861 and 1865, would decrease in the new
structure and with a new surgeon. But although their new bricks, red or
not, failed to deter microbes, their new professor did. Convinced that
fermentation—and not vapors—caused sepsis, Lister turned to carbolic
acid which had proved useful in cleaning sewers. Using carbolic acid as
disinfectant, he was able to reduce surgical mortality in his male accident
ward from forty-five to fifteen percent between 1865 and 1869. Lister's
demonstration of antisepsis in the clinic preceded his success as a microbe
hunter by several years; it was not until 1873 that he validated the "germ
theory of fermentation" by isolating the first pure culture of a microbe, the
lactobacillus. Lister, who was the link between the sanitarian and bacteriologic revolutions, paid belated homage to Semmelweis as his colleague in
antisepsis; "I think with the greatest admiration of him and his achievement
and it fills me with joy that at last he is given the respect due to him."

NIGHTINGALE AND PASTEUR

But in the largest lying-in hospital of Europe puerperal sepsis remained
unchecked as if Gordon, Holmes, and Semmelweis had never lived. Between April 1 and May 10, 1856, the Maternité of Paris experienced 64
fatalities out of 347 confinements. The hospital had to be closed and the
survivors took refuge at the Hôpital Lariboisière, where they nearly all
succumbed, seemingly pursued by "pestilent vapors." Historians suggest
that the Maternité probably had the highest sustained mortality from
pyemia after birth of any large institution: between 1861 and 1864, almost
one-fifth of all deliveries resulted in the death of the mother. While those
conditions shocked many Anglo-Saxon medical visitors to Paris, they failed
to weaken English belief in the zymotic or atmospheric theory of contagion.

In a letter of May 19, 1862, Florence Nightingale replied to Dr. Charles
Shrimpton who asked her why more died of pyemia in Paris than in
London: "Those who know the construction of Paris dwellings know full
well the reason. Given a surgical patient admitted into a Paris hospital:
—there is strong ground for believing that patient labours under pyaemia
to begin with. As regards the general practice result, patients die more

frequently in Paris where the previous pyaemic condition is more manifest than in London where it is less manifest." The appeal to zymotic theory (i.e., previous pyemic condition) yields itself readily to social analysis. "Pyemic conditions" was English middle-class shorthand for dirty, lower-class, and foreign. Charles Rosenberg has pointed out that for Florence Nightingale and her followers, mid-nineteenth-century hospitals were places of disorder that middle-class doctors and nurses found somewhat menacing: "The new industrial cities seemed to be alive with a threatening and potentially chaotic working class—a class that produced a disproportionate number of victims of epidemic disease and provided virtually all the occupants of urban hospital beds. Zymotic images were ideal for describing both the city's and the hospital's fermenting and disorderly life—a below stairs without an above stairs to control it."

At the Maternité things went from bad to worse as doctors tried to exorcise the "epidemic genii." In 1864, there were 310 deaths out of 1,350 confinements and in 1865 the hospital was again closed. When it reopened at the beginning of 1866, Dr. Trélat, the surgeon-in-chief, complained that "the sanitary condition seemed perturbed"; by February, there were again 28 deaths out of 103 births.

And so the pestilence continued. While the sanitarians may have proven their case in Boston, Budapest, and Glasgow, women continued to die in Paris until the little army of microbes was identified as the cause. It took the master of microbe hunters to spot the exact foe. Louis Pasteur, having preserved the wine, silk, and dairy industries of France, turned his attention to puerperal sepsis. At the Academy of Medicine in 1879, Pasteur interrupted a discussion of puerperal fever epidemics, rising from his seat to insist: "None of those things cause the epidemic; it is the nursing and medical staff who carry the microbe from an infected woman to a healthy one." As one clinician replied from the podium that he doubted the germ would ever be found, Pasteur rushed to the blackboard to draw several grapelike chains of microbes, and pointed: "There, that is what it is like!" That celebrated demonstration, and the experimental work on which it was based, was soon published to great acclaim; Pasteur's discovery that puerperal fever is caused by the streptococcus was a turning point in the bacteriological revolution.

Charcot's memoir of Pasteur (1895) showed the temper of the time: "Obstetrics has been revolutionized and puerperal infection no longer transforms the lying-in maternity hospital into a huge necropolis. On this

point, the testimony of the famous English surgeon, Lister, is most significant and authoritative... in 1892, when bringing to Pasteur an address from
the Royal London Society [*sic*], he exclaimed: 'really, there is not in the
whole world a man to whom medical science owes more than it does to
you.'"

SUMMARY

For much of the nineteenth century many thousands of women died of
"an almost, if not entirely preventable" disease, because public beliefs—in
the zymotic theory, or in the benign agency of doctors and midwives—prevailed over the facts of public health. In 1996, only three of every 100,000
women in the industrialized world died of puerperal sepsis, a rate that
would have astounded the sanitarians and microbe hunters of yesteryear.
But in countries that lack sanitation, antibiotics, or even an organized social
structure (Chad, Nepal, Somalia), the rates were higher by one hundred-
fold. Changes in the public health of those lands, anthropologists are quick
to point out, would require major revisions of public belief. Those in the
West, however, who place public health first among the values of civilization have little reason to feel smug. In the United States, for example, many
thousands suffer or die from infectious diseases such as Lyme disease,
ehrlichiosis, rabies, tuberculosis, or AIDS—afflictions that are "almost, if
not entirely preventable." A sanitarian of the future might argue that our
public beliefs in animal rights, civil liberties, and the bottom line have
prevailed over public health at the end of the twentieth century; he might
do worse than to raise the voice of Holmes:

> *And lo! the starry folds reveal*
> *The blazoned truth we hold so dear:*
> *To guard is better than to heal*
> *The shield is nobler than the spear*

5

BÊTE NOIRE

. . . Suppose we dance, suppose we run away
into the street, or the underground
he'd come with us. It's his day.
Don't kiss me. Don't put your arm round
and touch the beast on my back . . .

Yes, I too have a particular monster
a toad or a worm curled in the belly
stirring, eating at times I cannot foretell, he
is the thing I can admit only once to
anyone, never to those who have not their own.

Never to those who are happy, whose easy language
I speak well, though with a stranger's accent.

KEITH DOUGLAS, "Bête Noire" (1944)

"Bête Noire" is set in Piccadilly during the long winter between Alamein and the Normandy invasion. At the time, Douglas had pretty much recovered from wounds inflicted by German 88's in the Western Desert and by spring he was back in action. On June 9, three days after landing in France, he was killed behind enemy lines. He was twenty-four years old and that posthumous poem can be read as his plea for a pardon. But there's no need for pardon, we want to tell him, we know about the bête. We can track its spoor to a brief memoir, *From Alamein to Zem Zem*, in which he describes his escape from a blasted tank over a minefield of wrecked armor and oil-stained corpses:

Presently I saw two men crawling on the ground . . . I recognized one as Robin. His left foot was smashed to pulp, mingled with the remainder of a boot. But as I spoke to Robin saying, "Have you got a tourniquet, Robin," and he answered apologetically, "I'm afraid I haven't, Peter." I looked at the second man. Only his clothes distinguished him as a human being, and they were badly charred. His face was gone: in place of it was a huge yellow vegetable. The eyes blinked in it, eyes without lashes, and a grotesque huge mouth dribbled and moaned like a child exhausted with crying.

For most combatants, the numbing effects of such battlefield nightmares are relatively short-lived, for some they last a lifetime. Those more permanently affected have been said to be "touched by fire" (the American Civil War), suffering from "shell shock" (World War I), or afflicted by "traumatic neurosis" (World War II). Curiously, in each instance the overt symptoms that patients displayed seemed to split along class lines. "Officers complain of nightmares and bellyaches, enlisted men think they've been paralyzed," our psychiatry instructors told us at Fort Sam Houston after the Korean War.

The syndrome became codified after that wretched war in Vietnam. The beast on the back—and a grab bag of other distressing symptoms—came to be called the "posttraumatic stress disorder," or PTSD. According to the American Psychiatric Association, the official definition of this condition requires a traumatic event "outside the range of usual human experience . . . one that would be markedly distressing to almost anyone," followed by such symptoms as the repetitive recall of the trauma in dreams or feelings, psychological numbing, amnesia, insomnia, or other forms of autonomic arousal. Readers of Robert Graves, Siegfried Sassoon—or Pat Barker, for that matter—should not be surprised that this description of PTSD turns out to resemble in good part the description of shell shock that has become part of the modern literary tradition; the Yankee psychiatrists are, after all, simply describing the same beast on different backs.

It might be argued that the beast has been there since records were kept; nightmares have been with us always. Among physicians, Hippocrates had the first look: "but the worst of all is to get no sleep either night or day; for it follows from this symptom that the insomnolency is connected with sorrow and pain." Two better-known passages from Shakespeare suggest that the bard had also met PTSD; a military example is that of Hotspur, who with beads of sweat on his brow rolls in sleep restlessly to mutter "Of prisoner's ransom and of soldiers slain/And all the currents of a heady flight" (*Henry IV*, pt 1, act 2). Guilt after mischief leads the Macbeths to "eat our meal in fear and sleep/In the affliction of these terrible dreams/That shake us nightly: better be with the dead/Whom we, to gain our peace, have sent to peace" (act 3, scene 2). Moving down a notch or two, Rudyard Kipling joined military to civilian motifs and added the element of class in "Gentlemen Rankers":

> *If the home we never write to, and the oaths we never keep,*
>   *And all we know most distant and most dear,*
> *Across the snoring barrack-room returns to break our sleep,*
>   *Can you blame us if we soak ourselves in beer?*
> *When the drunken comrade mutters and the great guard-lantern sputters*
>   *And the horror of our fall is written plain,*
> *Every secret, self-revealing on the aching white-washed ceiling.*
>   *Do you wonder that we drug ourselves from pain?*

Each of these three themes, battle trauma, internal guilt, and class resentment, is crisply handled in Allan Young's recent scholarly study, *The Harmony of Illusions: Inventing Post-Traumatic Stress Disorder* (1995), of how the syndrome of PTSD was "invented." But he would disagree with the notion that it has always been with us.

> The traumatic memory is a man-made object. It originates in the scientific and clinical discourses of the nineteenth century; before that time there is unhappiness, despair, and disturbing recollections, but no traumatic memory, in the sense that we know it today.

Young develops this argument as a lucid case study of how medicine and society have for over a century and a half managed to construct this "man-made" disorder, reinforcing social history with an expert piece of field work. A medical anthropologist at McGill University, Young describes his observations at the National Center for the Treatment of War Related Post-Traumatic Stress Disorder (*sic*) somewhere in the Midwest of the United States. He visited the Veterans Administration center thirteen years after the troops left Vietnam; the transcripts convince us that for many of those vets the battle rages still.

Among the VA patients there was no Hotspur, no Douglas, no Siegfried Sassoon. And as might be expected from the economics of VA medicine in the United States, there were certainly no gentlemen rankers. The VA housed only hard-core remnants of the conscript army we sent to Asia and despised as losers when they came home: the drifters, the violent, the addicted, the dim. Each carried his own beast on his back, which the therapists called the "pathogenic secret," waiting to be confessed in a climax of relief. But there were few epiphanies at the VA. In fact, writes Young, "There are occasional 'disclosures' of course, since this is what the clinical ideology demands, and some of these narratives are vivid and charged with emotion. But there are no real climaxes; there is no point at which every-thing—narrative, affect, and remission—seems to come together." Life at the center was rather monotonous "full of unending hours of talk, punctu-ated by the incessant drip-drop of tiny signifying moments." Some of these are on record:

> MARTIN: Well, you get orders to burn a village, and a gook tries to put the fire out while you're trying to burn his hootch. He fucks with you, and you show him that you can fuck with him. You can push him away, or you can kick his ass, or you can do what we usually did: you can shoot him.

> JACK: You can't see anything because of the smoke—awful smell—all kinds of shit burning. We entered the village, and there were bodies all over the place. Near me there was a dead old woman and a young girl—but the girl was alive. She'd lost one leg and was going around in a circle on the ground, crying out but not making any sound. The marine next to me takes out a hand gun and shoots her in the head. I was completely pissed by the fucking thing, all of it. All I wanted to do was trash people and that's what I did. I didn't care who they were, and I'd just as well have killed U.S.

> PETER: In Vietnam, we didn't have an objective. We weren't allowed to accomplish anything. They just sent people there to fart around and to die.

Those are not the voices of Douglas or Kipling or Siegfried Sassoon; they're the sounds of rats in a trap.

Psychiatrists, psychologists, epidemiologists, and military folk have long tried to understand why such responses to traumatic memories often fall into a stereotypic pattern while others vary widely with time and place. One can pinpoint the birth of traumatic memory to a machine-made disaster, the train wreck. John Erichsen was perhaps the first physician to describe the syndrome during the 1860s while examining victims of British railway accidents, attributing the syndrome, which he called "railway spine," to vaguely defined neurological mechanisms that originated in dorsal trauma. "Spinal irritation" was a widely popular diagnosis at the time: Henry and William James acquired the disorder to sit out, as it were, the American Civil War. Their plucky sister Alice spent a lifetime in bed on account of her spinal affliction, the "dorsal trouble in the blood" which William believed to run in the family. The disorder was based on Marshall Hall's earlier description of the reflex arc, a reduction of our higher mental functions to local electrical circuits, a now archaic, if persistent, belief. The notion that our spines harbor the secret of health and disease remains at the root, literally, of such curious American practices as osteopathy, chiropractic, and Christian Science.

The continental alienists, Charcot, Janet, and Freud, suggested that the syndrome need not be produced by true physical trauma; psychological wounds were sufficient. It is therefore appropriate that Freud's heuristic theory of the dynamic unconscious dominates the study of traumatic memory today. But, in the course of the Great War, the locus of the disorder shifted from Erichsen's railway spine and Freud's consulting room hysteria to the battlefield, where large numbers of soldiers were now given the diagnosis of shell shock. Young tells us that:

> A half-century after the publication of Erichsen's first book on railway accidents, physicians serving in the Royal Army Medical Corps, like their counterparts in other combatant armies, were witnesses to an epidemic of traumatic paralyses, contractures, anesthesias, and abulias. It was as if a hundred colossal railway smashups were taking place every day, for four years. By war's end, 80,000 cases of shell shock had been treated in RAMC medical units, and 30,000 troops diagnosed with nervous trauma had been evacuated to British hospitals. After the war, 200,000 ex-servicemen received pensions for nervous disorders.

It was during this epidemic that the pioneering W. H. R. Rivers—also a medical anthropologist—introduced the Freudian dynamic unconscious to the treatment of shell shock. He is one of the two heroes of Pat Barker's splendid trilogy (*Regeneration, The Eye in the Door*, and *The Ghost Road*) and Barker shows an uncanny, or at least novelistic, feel for his clinical touch. Freudian themes merge with Jungian anthropology in her trilogy—as they did in Rivers' own life. But because—or despite—of his Jungian leanings, Rivers was an early convert to the Freudian cause. In a 1917 (!) essay for

Lancet entitled "Freud's Theory of the Unconscious," he paid homage to his wartime enemy and professional colleague:

> Instead of advising repression and assisting it by drugs, suggestion, or hypnotism, we should lead the patient to resolutely face the situation provided by [the patient's] painful experience. We should point out to him that such experience . . . can never be thrust wholly out of his life. . . . His experience should be talked over in all of its bearings.

Much of Pat Barker's trilogy is based on Rivers "talking over" the lives of Siegfried Sassoon, Robert Graves, and those others who had walked the ghost road. Nevertheless, Rivers differed with Freud on several points, arguing that the shell shock cases he treated could in no reasonable sense be linked to replays of childhood sexual theatrics. But by and large Freud's patient unearthing of the past was preferable to the heroic countershock treatments then fashionable in the military. The talking cures of Freud and Rivers, based on restoring the mind's equilibrium, were—consciously or not—in keeping with the physiological principles of homeostasis enunciated by Walter B. Cannon of Harvard (1871–1945), the Claude Bernard of American physiology. Cannon, a student of physiologic shock, put neuroendocrinology on the map forever in his "The Emergency Functions of the Adrenal Medulla in Pain and the Major Emotions":

> The organism which with the aid of increased adrenal secretion can best muster its energies, can best call forth sugar to supply the laboring muscles, can best lessen fatigue, and can best send blood to the parts essential *in the run or the fight for life*, is most likely to survive. [my italics]

And so flight or fight it was, and traumatic memory danced to the tune of adrenaline.

Interest in the syndrome declined after World War I but revived during the 1940s, when Abram Kardiner, an American psychoanalyst who had treated traumatized veterans in the 1920s, codified the criteria for its diagnosis and distinguished its delayed and chronic forms (Macbeth v. Hotspur, one might say). We owe to him the terminology of "traumatic neurosis"— and I remember distinguishing its various forms in order to define disability claims as we mustered its real and alleged victims out of service. The traumatic neuroses of World War II were considered diseases of adaptation to relentless stress, a notion again in concord with physiologic theories of the time. Hans Selye (1907–1987) put both "adaptation" and "stress" on the medical map, basing his work on the other major secretion of the adrenal, cortisone. In a short note to *Nature* (1936), Selye correctly identified cortisone as the stress hormone:

> We consider the first stage [of response to sublethal injuries] to be the expression of a general alarm of the organism when suddenly confronted with a critical situation and therefore term it the 'general alarm reaction.' Since the syndrome as a whole seems to represent a generalized effort of the organism to adapt itself to new conditions, it might be termed the 'general adaptation syndrome.' It might be compared to other general defense actions such as inflammation or the formation of immune bodies.

And soon thereafter, disorders caused by physical or mental stress, like those provoked by noxious microbes, were viewed as flaws of mental adaptation, an inflammation of the spirit gone awry, a splinter of the soul. "I'm all stressed out" we say today, à la Selye.

The field changed radically when the nomenclature was revised in consequence of the American experience in Vietnam: PTSD was born in the rice paddies. But the diagnosis achieved general acceptance only in 1980, when PTSD was included in the third edition of the American Psychiatric Association's *Diagnostic and Statistical Manual of Mental Disorders* (DSM-III). Young's book details the "political struggle waged by psychiatric workers and activists on behalf of the large number of Vietnam War veterans who were then suffering the undiagnosed psychological effects of war-related trauma." American psychiatrists of widely different persuasions and with divergent agendas played a role in this struggle, high praise is due a cast of effective doctors who hammered out the names we still use today for the many forms of mental disease. One might credit such pioneers of psychiatric nosology as Robert Spitzer, Gerald Klerman, and Myrna Weissman, while Robert Jay Lifton and Nancy Andraesen get full marks for squeezing PTSD into the canon. And when the latest revision of the code (DSM-IV, 1994) made room for all the lost Martins and Peters and Jacks, Young concluded

> that the publication of *DSM-IV* is a signifying moment. It signals the repatriation of the traumatic memory, the act of bringing it back home from the jungles and highlands of Vietnam. The collective memory of their war dims and gradually merges with the memories of older, half-remembered wars fought in Korea, Europe, and the Pacific. As the veterans of Vietnam age and fade, and their patrons in government adopt new priorities, a chapter in the history of the traumatic memory draws to a close.

Some would judge that the story of PTSD—or Pat Barker's trilogy in which Rivers turns the syndrome around—supports the argument that medicine and society interact to frame or to "construct" diseases. That school of social thought teaches that the facts of science are contingent on their context. Young, who is clearly of that school, agrees that "the suffering is real; PTSD is real" but goes on to ask, "can one also say that the facts now attached to PTSD are *true* (timeless) as well as real?" Probably not, one might reply, but even at this stage in the history of mental science, there are *some* facts that already bid fair to become as timeless as the law of perfect gases. $PV = nRT$ is not more of a fact than that general paresis is due to the spirochete of syphilis, that Alzheimer's disease is associated with neurofibrillary tangles in the brain, or that lead drove the hatter mad. Medical science is already in great debt to social scientists (the names of Robert Merton, Edward Shorter, and Richard Titmuss come immediately to mind), but if psychiatry is too important to be left to doctors, it is certainly too important to be left to today's social science. My colleagues in molecular biology, proud that they have just completed the first projection of the human genome (*Nature*, 14 March 1996), look at psychiatrists and social scientists in much the same way that the early physicists looked at the alchemists: as funny men in conical hats who melt base metals in a pot. No matter. I'm inclined to agree with Lewis Thomas that, with time and hard work,

> a solid discipline of human science will have come into existence, hard as quantum physics, filled with deep insights, plagued as physics still is by ambiguities but with new rules and new ways of getting things done. . . . If anything like this does turn out we will be looking back at today's social scientists, and their close colleagues the humanists, as having launched the new science in a way not all that different from the accomplishment of the old alchemists, by simply working on the problem—this time, the fundamental, primal universality of the human mind.

6

SCIENCE FICTIONS

He had been awarded the Rumford Medal of the Royal Society. Several of his researches, so I am informed, on good authority, are classical beyond dispute. That is, they have been passed by time. The suggestion is now . . . that he went in for cooking his results.

C. P. SNOW, *The Affair* (1960)

Beneficentiam et veracitatem est
[It is beneficial and verifiable]

DIONYSIUS LONGINUS, *de Sublimi* (1st century A.D.)

From Longinus to C. P. Snow, educated folk believed that great works of art or science had a life of their own, that they had been "passed by time." They transcended epoch, creed, or gender and were constructed in, but not by, society. Masterworks of art or the laws of science were considered to have a life of their own away from native soil: they came value-free, as we say these days. Each observer, each time, would have the same, reproducible response to a work of genius, be it the *Nike of Samothrace* or the Bragg equation of X-ray diffraction ($n\lambda = 2d \sin\theta$). Longinus knew that the sublime could be the source of both subjective pleasure (*beneficentiam*) and objective standards (*veracitatem*).

Alas! Few who move in the culture of the arts today still hold this belief; we are told that what is beautiful and sublime is socially constructed and that art cannot be value-free. Indeed, the notion is suspect that any work of art can approach the sublime, and the merely beautiful is dismissed. Alas! once more: a broad movement is afoot in our academies to extend this view to science as well. All science is socially constructed, we are told. Science is as society does; science can never free itself from the values of its time.

Half a century ago, the notion that there is no such thing as value-free science was championed mainly by National Socialists or Stalinists of the Lysenko School. In the Third Reich, science was shaped by one's *Weltanschauung*. But nowadays, half a century after the purges of "Jewish physics" and "capitalist genetics," the social-construct camp has attracted a new crop of more respectable devotees. Feminists, proponents of diversity, pleaders for alternative healing, and what appears to be a majority of new historians, philosophers, and sociologists of science believe that science, by definition, is the construct of those who do it. They also believe that science is pursued in the interest of those who pay for it, an establishment of white, male Europeans.

There is no such thing as a scientific truth, they tell us, but only a temporary construct that society permits to *be* true: objective science is a fiction. It should not surprise us, therefore, to be told by sociologists that there is a uniquely feminist way of understanding genetics, by historians that there are distinct Afro-American or Jewish insights into cell biology, or by enthusiasts of alternative medicine that there is a traditional Chinese treatment for every occidental tourist. Once again, it all depends on your *Weltanschauung*.

CONSTRUCTING SCIENCE

The most frontal assault on value-free science has been launched by feminist critics who attack the masculine bias of Western medical science. Male science, we are told, is a kind of science fiction and women will change the way it is done. Feminists argue that gender is a powerful determinant of how science is done, written, and rewarded. Evelyn Fox Keller has made the perfectly correct observation that much of Western science has been expressed as "the rhetoric of domination, coercion, and mastery." She goes on to complain that these Freudian, erotic elements "persist throughout history only *sotto voce* [and] are expressed as a kind of primal scene of Western science: masculine science having its way with Mother, or Virgin, Nature."

Nature, Fox Keller assures us, is everywhere depicted as female in myth and legend, art and song—and in scientific writing. Since the masculine, aggressive view is counter to the female view of cooperativity, sensibility, and a "feeling for the organism," women have been written out of power.

Medical science has suffered in consequence: for example, embryology texts tend to depict the egg as *passive* and the sperm *active* and therefore to understate the kinetic contribution of the egg's membrane to entry of sperm. And the written history of DNA has neglected its debt to the special nurturing talents of one particular woman and her feeling for the organism. Nobelist Barbara McClintock's discovery of transposable elements in maize genes has been downplayed in favor of all those male laureates who took apart genes like so many Tinkertoys.

Such arguments—sotto voce or prima facie—are of course a subset of social-construct theories. But as is so often the case with anecdotal sociology, they do not ask "compared to what?" One might remind Fox Keller that not only nature is traditionally depicted as female in the visual arts: so are medicine, biology, physics, research—and science herself. The owl of Minerva wasn't hiding in a man's skirt. In the German and French languages, for example, the sciences are all feminine nouns: *médecin*, a doctor, is male all right, but *médecine*, what he practices, is female again. Ditto for the German *Mediziner and Medizin*.

Moreover, when it comes to the rhetoric of sperm and egg, feminists have neglected a field of study that put Jacques Loeb on the front pages of American newspapers at the turn of the century. He showed that eggs develop into embryos when activated by simple salt solutions or chemicals, all without the help of a single sperm: parthenogenesis in a dish. Since the days of Loeb, developmental biologists, both male and female, have documented that the egg is as active as any cell under the sun, or for that Freudian matter, the sun itself. Finally, there is that "feminine" transposition of McClintock's maize genes: it does not help the cause of women in science to be told that a "feeling for the organism" is best developed by a lifelong devotion to the ears of corn. One might, in fact, point out that transposable elements in the immunoglobulin genes of *humans* were described by male scientists—Dreyer and Bennet (1965) in theory, and Hozumi and Tonegawa (1976) in fact—independently of anything Barbara McClintock ever wrote. I would argue that some men and some women of science have a feeling for the organism, others not.

FICTIONAL SCIENCE

Of the many arguments in favor of gender-, class-, and value-free science, the most obvious one is the general applicability of physical laws, under any regime, in every hemisphere, here on earth and there in space. That Bragg equation, for example, applies as neatly to early metal fatigue in space station *Mir* as to that fudged photograph in C. P. Snow's *The Affair*. East is East and West is West, but the twain who shall never meet in Kipling have found common ground at $n\lambda = 2d \sin\theta$.

But, as C. P. Snow told us a generation ago, perhaps the chief argument for "true" science is the persistence of the false: fraud. For if there is no such thing as absolute scientific truth, there can only be relative scientific fraud. If the faker plays only a social role, he does not commit fraud when he cooks the books, he enacts a social drama. If there is no external, objective standard out there, no there there—in Gertrude Stein's phrase—how can one fudge? A student of scientific fraud, Marcel C. La Follette, points out "the question of 'how much fraud is there' represents an inherently unanswerable question as long as it is only posed in opposition to something labelled good science.'" I should hope that the question remains unanswered, because "good science," by virtue of its reproducibility, is less at risk from fraud than from those who would police it. For while scholars talk of social construction, our rulers talk of politics.

One example of the shambles that result from the collision of science and politics is the dramatic confrontation between Rep. John Dingell (D.-Mich.) and David Baltimore, a Nobel laureate, over alleged fraud by his immunological collaborator, Dr. Imanishi-Kari. The story of those sorry events in the late eighties could thrill only the true policy wonk. Who would have believed that a paper on immunoglobulin genes would come under scrutiny by Congress—and the Secret Service would have been called in to check notebooks in Boston? La Follette describes the temper of the times.

> The atmosphere in these hearings has been charged. The rooms are packed. Whistle-blowers and the supporters of the accused jostle for seats with a blossoming [sic] number of professional staff members conducting studies at federal agencies and scientific academies. Members of the press wear guarded but skeptical expressions. And everyone exchanges knowing glances, trying pretentiously to look like insiders.

It was in this sort of atmosphere that David Baltimore blurted out to the congressman that he had more important things to do than to worry over experiments that Dr. Imanishi-Kari may or may not have fudged. An unbiased observer might say that Baltimore was right, for in the course of the sad, prolonged, and essentially trivial proceedings that have persisted since the fraud was alleged in 1986, Baltimore's labs at Rockefeller and MIT made a major discovery: a factor called $NF_\kappa B$ that controls the gene machinery of host defense. This discovery may indeed be of greater importance than the reverse transcriptase for which he won the Nobel prize. And the subsequent careers of Drs. Baltimore and Imanishi-Kari seem to have vindicated his claim. He is now president of the California Institute of Technology, and Dr. Imanishi-Kari is hard at work in the lab.

Most accounts of fictional science are written by those who favor social construct theories rather than those who believe that science is value-free. In many of the recent scandals, the ones who have been directly implicated in cooking the data are those doing the scutwork of science at the bench, on the wards, or in the stacks. The fabrications have been traced to the ambitious postdoc, the assistant, the instructor, the foreigner, the woman. Imanishi-Kari has herself complained that they singled her out of the Baltimore group because she was a foreigner and a woman, and "they got at [Robert] Gallo through an Eastern European." Those who have been "gotten at" are some of the major stars of the scientific and medical firmament and include not only David Baltimore (MIT, Rockefeller) but also—among others—Fritz Lipmann (Harvard, Rockefeller), Robert Good (Memorial Sloan-Kettering), Ellis Reinhertz (Harvard), Eugene Braunwald (Harvard), Philip Bondi (Yale), Philip Felig (Yale), and Robert Gallo (NIH). The misconduct of which their underlings were accused is a catalogue of rash or ludicrous behavior— of planting false radioactive bands in acrylamide gels, of dropping reagents into test tubes at night, of mislabeling photos, of inventing patient records, of diddling with electrocardiograms, of imagining new diseases or painting black skin grafts on little white mice. Each case is a sorry tangle of Freudian dynamics in the thick underbrush of the student/mentor forest, where—to continue the metaphor—many a mighty oak has fallen when the matter comes to light.

But, fictional science is no real threat to factual science. The point is that no matter who fudged what, where it was published, or what the motives were of those who perpetrated the fraud, it remains the virtue—the arete— of science that fraud cannot go undetected for long. If the substance of a scientific or medical claim is important enough, it will be replicated and if it is trivial, if the experiment is not a rigorous test of a hypothesis or the description of a novel phenomenon, it will join the bulk of the literature in remaining unread, unquoted, and forgotten.

In that sense, another geographic metaphor may be useful. Ecologists complained that the 1993 floods of the Mississippi river would pose a greater danger for riverbank agriculture than ever before because the many dikes, levees, and deviations of modern flood control have interfered with the self-cleansing power of the river. Like the Mississippi, Western science since the Great Instauration has been self-cleansing by nature. Unlike the river, science has steadily increased its capacity to cast off muck and debris, flotsam and jetsam, error and fraud. Western science after Bacon may be described as a pretty good method for seeing what floats.

The self-cleansing aspect of science, its capacity to reinvent itself, dictates that any effort to fudge, fake, cook, or steal the data is, literally, superfluous. It follows that the motive of those who commit fraud in science cannot be other than destructive, whether of self (always) or of superiors (often). The unconscious, dynamic aspects of these shenanigans cannot be addressed short of intensive training in psychopathology. No ordinary scientist should be expected to discern what inner urges lurk in the hearts of the few who fudge.

"Theft is theft and raid is raid, though reciprocally made," wrote Robert Graves in another context. Theft, raid, fraud, lying, and cheating deserve punishment, but I doubt that the introduction of preemptive antifraud measures in lab or in clinic will lessen the incidence of snitching. And to finish the aqueous metaphor, the efforts of scientific whistle-blowers or vigilantes such as Ned Feder and Walter W. Stewart—federal scientists who have pursued Baltimore, Braunwald, and others—are only likely to muddy the waters. The remedy probably lies in vigilance rather than vigilantism (a phrase of Berkeley's Howard Shachman).

Indeed, although scientists go about their work with a number of goals in mind, their real worth is measured in reputation—and reputation is guaranteed only by the repeatability of their work; *veracitatem*, one might say. "Do you have that latest reprint?" is the question about a hot paper. And reprints that *stay* hot are what make reputations. In rephrasing Longfellow's "Psalm of Life," Sir Hans Kornberg assures us that:

> *Lives of great men all remind us*
> *We can make our lives sublime*
> *And departing leave behind us*
> *Reprints on the sands of time.*

Fraud therefore implies its opposite, truth, which in rhetoric has another opposite: fiction. Perhaps the reason that stories like the Baltimore case make national news is because they become test cases of what is factual and what is fictional science. They resurrect our deepest worry that today's major discovery will be tomorrow's egregious error. They also reinforce the prejudices of the unlettered, the sentimental, and the fans of revealed religion that science is just another self-promoting racket, that its truths are simply convenient fictions.

SCIENCE AS WRITING

Nevertheless, fictional science is only one form of science fiction. Literary folk have now come up with their own, unique contribution to the value-free/social-construct debate. Calling themselves "countertraditionalists," they argue that science and literature are pretty much alike—i.e., socially constructed—since they are expressed as "text" and therefore subject to analysis by the instruments of literary criticism. David Locke, who is both a published chemist and professor of English, points out in *Science as Writing* (1992) that the traditional view of value-free science neglects the textual aspect of its transmission and that the

> countertraditionalists [i.e., Locke and friends] no longer see the scientific document as the bare, unadorned record of what the scientist has found writ in nature; rather, they see it as an artifact shaped by the collective activities of scientists and employed by them in their efforts to reach consensus.

This restatement of the *denkkollektif* theory of Ludwik Fleck (see Chapter 10) is made more, if not entirely, persuasive by Locke's examination of how science and literature use similar rhetorical devices. Locke tries to persuade us that since the compositional processes, if not the results, of science and literature are akin, they must both be socially shaped. But—to be fair—he is also clear as to how science and literature differ in the use of their own history. A. N. Whitehead told us that a science that hesitates to forget its history is lost, and Locke explains "why learning a science and learning the current language of the science go hand in hand. This is . . . why scientists are so uninterested in the historical documents of their sciences; those documents are written in languages that scientists no longer speak."

There is no question but that many of the crucial "texts" of science (e.g., Watson and Crick's famous report of DNA structure in *Nature*, Robert B. Woodward's chemistry papers, or Albert Einstein on relativity) can be subject to the various readings that form the repertoire of literary criticism. Unfortunately, when the countertraditionalists get going their jabber sounds more like a David Lodge parody than the David Locke original:

> In the argument, we shall adopt the convenient fiction that six discrete theories exist, each making its own competing claim for validity in the reading of the text: (1) representation theory, which sees the literary work as essentially a representation of the real world; (2) expression theory, which views that work as an expression of its author's thoughts and feelings; (3) evocation theory, which values it as an evoker of response from its readers; (4) art-object theory, which judges the work as an object d'art, interesting in its purely formal properties; (5) artifact theory, which situates the work in its social milieu; and (6) instrumentality theory, which places the work among the signifying systems that organize, structure, indeed constitute, the world.

Jargon aside, are any of these theories valid for "a reading of the text"? Locke applies lit/crit to the famous ending of the Watson–Crick paper: "it has not escaped our notice that the specific pairing [of bases in DNA] we have postulated immediately suggests a possible copying mechanism for the genetic material." Locke argues that just because this report has scientific meaning it is not devoid of interest as language:

> Does this mean that we are viewing this scientific report as an art object, as the art-object school of criticism views a poem? Essentially, yes. I do not intend to say, obviously, that this document is a poem, but merely that its language is of interest just as is the language of a poem, *as language*.

But this sort of analysis tells us only that the tools of literary criticism are sufficient to cope with any conceivable text. Since it does not compare this particular text (Watson and Crick) to the two papers on DNA structure that accompany it in the same issue of *Nature* (Wilkins, Franklin *et al.*), it misses the point that not only the rhetoric of Watson and Crick's paper (a fiction) but also its unique attention to base-pairing as an element common to structure *and* function of DNA (a fact) makes this paper transcend mere eloquence. Watson and Crick based their solution of DNA structure on the Bragg equation. Is $n\lambda = 2d \sin\theta$ an art object? Is it a text? Is it rhetoric? Or is it a fact?

Unpersuaded by the "countertradition," I would maintain that the helical structure of DNA or the Bragg equation will stand forever as facts of nature, value-free, when the societies that constructed them—or when language itself—will have been forgotten.

SCIENCE AS LITERATURE

Another way to slip hard science into the soft shoes of social construction is to blur the boundaries between science and literature. A good example of the genre is Lawrence Rothfield's *Vital Signs* (1992), a work which claims to "have treated medicine almost as a species of literature, and certainly as an art, a set of techniques that the doctor uses to represent illness." But, simply because doctors often tell stories does not mean that they are storytellers. The countertradition, indeed!

Rothfield documents how one genre (the realism of Balzac and Flaubert) was displaced by another genre (the naturalism of Zola) and correlates "the change in discourse" of the novel by relating it to the displacement of one form of scientific thought (the clinical medicine of Xavier Bichat) by another (the experimental medicine of Claude Bernard). Since the works of Bichat and of Bernard feature prominently in the novels of Flaubert and Zola, respectively, the connection is appropriate. Rothfield extends his analyses of nineteenth-century novels and medical science to include works that range from Balzac to George Eliot, to Henry James, and to Conan Doyle. In the course of this task, he unearths an archaeology of medical science which expands the reach of the new historicism. He claims, correctly, I believe, that:

> the history relevant to literature includes the history of science, that the sciences are a cultural phenomenon providing part of the cultural basis for literature just as other kinds of intellectual activity do. This is not a very daring suggestion, to be sure. In thinking of medicine in particular as a culturally implicated rather than purely scientific practice, I am by no means alone. But the archeological method I have used to analyze the cultural resonance of this practice distinguishes my book in several ways from other recent cultural studies focusing on medicine.

There can be no doubt that early nineteenth century notions of hysteria may be the intertext of *Madame Bovary*. Rothfield traces this theme to the family romance as Flaubert *fils* responds to his successful physician-father:

> In addition to its simplicity, this interpretation of the father: son relationship has another advantage ... it accounts for not only two but three

generations of medical genealogy in Flaubert's life as well as in his text. If the first medical generation is that of Bichat (recall Flaubert's description of Lariviére [the consultant called to Emma's deathbed], as one of a great line of surgeons that sprang from Bichat), the second generation is that of Flaubert's father and Lariviére, both of whom studied under Bichat. The third generation then belongs to Flaubert and . . . surprisingly enough, to Charles Bovary. No wonder Flaubert says that Lariviére's kind of surgeon is now extinct. Neither Charles nor Flaubert is a successful physician. In that sense, Charles represents Flaubert's failed ambition to become a doctor—and indeed one can trace many of the signifiers of failure borne by Charles (stuttering, falling into stupors, and so on) back to Flaubert, the idiot of the family. . . . Flaubert makes himself the true heir to Bichat's anatomical insights.

Not only Freudian, but economic lenses can be used to peer at the text of a great novel, or a great anatomical insight, for that matter. Karl Polanyi has described the "great transformation" of 1790–1840, which—among far greater consequences—caused fundamental changes in the "social construction" of the medical and writing professions. Rothfield uses this paradigm to unite the workaday lives of doctor and novelist: the great transformation resulted in "a shift—both for medical men and for novelists—from a relatively small and divided market to a unified mass market for their services." But while the marketplace of doctor and novelist may have shifted in similar fashion, medical science differs from novel writing as much as, shall we say, fact does from fiction.

These books of the countertradition persuade me that lit/crit analyses cannot be used to support the social-construct theory of science. Medicine is not "almost a species of literature" any more than the structure of DNA is a poem. Unlike the propositions of science, neither poems nor novels can be falsified since they are fictions already. Nevertheless, the "countertradition" which tries to intertwine the texts of art and science does have it right in one sense. Scientists and poets can both make asses of themselves; they can both hurt us. In his "Ode to Terminus" W. H. Auden called on the Roman god of boundaries to save humankind from the excesses of scientist and poet alike, introducing the poem with a couplet from Goethe's *Faust*:

> *In der Beschränkung zeigt sich erst der Meister*
> *Und das Gezetz nur kann uns Freiheit geben*
>
> [It is constraint that first reveals the master
> And only law can make us free]

As we've learned from schools of science that are *all* socially constructed, I speak of Lysenko and the Nazi geneticists, the laws of science are the best constraints we have on our wildest fancies. When we are asked to blur the difference between fact and fancy, science and fiction, we had better start looking out for our freedom. Bragg, with his $n\lambda = 2d\sin\theta$, had a better shot at uniting what is beneficial with what is verifiable than did the social constructionists, and $n\lambda = 2d\sin\theta$ is the turf we had better defend if we don't want to go way of the classical sublime.

7

CALL ME MADAME

Le concret c'est de l'abstrait rendu familier par l'usage
[The concrete is the abstract made familiar by usage]

PAUL LANGEVIN (1923)

On April 19, 1906, the forty-seven-year-old Nobel laureate, Pierre Curie, was run over by an oversize, horse-drawn wagon filled with bales of army uniforms. He was negotiating that tricky Parisian intersection where traffic from the Rue Dauphine, the Quai Conti, the Quai des Grand Augustins, and the Pont Neuf have created Gallic havoc for over a century. Curie had just quit a meeting of reform-minded university professors where he argued for legislation to improve the lot of junior faculty and to prevent laboratory accidents. He had planned to stop at his publisher's office on the Quai, but the office was shut because of a strike by equally reform-minded trade unionists. Absent-minded and somewhat radium-sick, he turned away in the spring rain, and was on his way to the library of the Institut when that six-ton wagon rumbled down the bridge from the Île de la Cité to crush his skull.

His death brought to an end two remarkably creative careers in physical science, his own and that of his wife, Maria Salomea Sklodowska—known to the world as Madame Curie. She later recollected that on the Rue Dauphine, "I lost my beloved Pierre, and with him all hope and all support for the rest of my life." She was right: for although Madame Curie was to survive her husband until 1934, her contributions to science after his death were less than innovative; she turned her tough mind to the application of their discoveries, to teaching young scientists, and to construction of the Radium Institute which she turned into a world center of physical science.

She also became a secular saint of feminist culture on both sides of the Atlantic, the subject of her daughter's hagiographic bestseller *Madame Curie* (1937), and heroine of sentimental publicists who hailed her for real or fancied radium cures. "More nonsense has been written about radium than the philosopher's stone," complained George Bernard Shaw in 1931, and he was right.

But there was no nonsense about the science. What a run the two Curies had together! In the course of six short years they had laid the foundations for the next century of physics and set the clock of our atomic age. That work earned Pierre and Marie Curie an acclaimed Nobel Prize in Physics (with Henri Becquerel in 1903) and Marie a more controversial Nobel Prize in Chemistry (1911). Those six years are the centerpiece not only of Eve Curie's biography of her mother, but also of all subsequent such works including those of Françoise Giroud (1986) or Susan Quinn (1995). Perhaps in keeping with the temperament of their subject, each is written with more diligence than grace.

The Curies were married in 1895, after Pierre had already become famous for his work with Jacques Curie on piezoelectricity (some crystals, e.g., ceramic or bone, generate an electric current when compressed). He soon earned his doctorate for studies with Paul Langevin on paramagnetic resonance (the moment of an atom or electron varies inversely with the temperature). It was the year that Roentgen took the first picture of the bones of his wife's hand by means of his novel rays.

By 1897, Henri Becquerel had found that uranium also produced rays—"emanations"—that left Roentgen-like shadows on photographic plates kept in the dark. Almost simultaneously, William Thomson, Lord Kelvin, discovered that the "ionizing" emanations from uranium imparted an electric charge to the air. In December of that year, Pierre and Marie set out to quantify the Becquerel emanations—ionizing radiation—of a great variety of natural substances. For this purpose they employed the piezo-electric quartz balance, an instrument that Pierre had designed, and by February had found that the residue of pitchblende from which uranium had been extracted gave far greater signals than uranium itself. They correctly deduced that there was an ionizing substance far more active than uranium lurking in the sticky brew. It was the same year that Zola wrote "J'accuse" and France split forever into the dreyfusards and their opponents.

By the end of 1898, the Curies had postulated that the new element, dubbed "radium," decayed into another which they called "polonium." They gave the name "radioactivity" to emanations from these elements. In 1902, by means of heroic preparative procedures, Marie Curie at last isolated radium in pure form. Later that year, Pierre calculated that one gram of radium emitted 3.7×10^{10} disintegrations per second: we call this amount of radioactivity one curie. And shortly thereafter he made the heuristic discovery that one gram of radium could heat one gram of water from 0 to 100°C: we call this sort of transformation "atomic energy" and nowadays it powers more than half of France. By 1903 Pierre and Marie Curie had won the Nobel prize: they had also come down with the first signs of radium sickness.

Six unmatched years of discovery in the setting of the Third Republic; axes drawn between right and left, church and state, theory and application, risk and benefit of a new science in a new century. It's a grand story and while the Curies are on the spoor of the new, with the Dreyfus case breaking about them, it's an exemplary tale of science in service to reason. But after Pierre's death on the Rue Dauphine, the story of Marie Curie becomes less a life in science, and more a story of The Career, The Scandal, and The Legend. Her biographers (hagiographers?) led by Eve Curie lead us through laundered accounts of widow Curie's adulterous affair with Paul Langevin, the outrageous attacks on her by the anti-Dreyfusard press, and the turns of her attention from science to the broader social scene. These proved to be as successful as her work in the lab. It was in recognition of the many mobile X-ray units she organized during the First World War that a grateful France forgave her for the Langevin affair by permitting her to establish the Radium Institute.

It is difficult to guess what inner doubts or conflicts might have troubled the pale, intense widow in a plain black dress who lived on the fashionable Quai de Béthune, the *entrepeneuse* who raised millions in France and the United States for her Institute by encouraging claims of cures for cancer. Nor does any material yet published yield insight into what must have been the remarkable relationship between Madame Curie and her daughter, a physicist at her mother's Institute, who married a brilliant young coworker to play out the story of *Marie et Pierre Redux*. Irène and Frédéric Joliot-Curie not only shared a Nobel prize for induced radioactivity in 1935—the third in one family—but also an abiding attachment to

Soviet communism. The story of the Curies reached from the Quai de Béthune to the podium of the Comintern.

The political and dynamic undertones of this part of the story are not in the public record. But that would have been just fine with Madame C: except for some painful letters addressed to her husband after his death, her private voice was as impersonal as her public speech. The Curies met in 1894 and were made for each other. Both Maria Sklodowska, daughter of a Polish gymnasium teacher, and Pierre, son of a Communard homeopath, were raised in the frugal folkways of the hardworking petite bourgeoisie. One catches the flavor of Curie's biographers—and a quaint view of genetics—from Susan Quinn's description of how Marie's mother learned to cobble her children's shoes: "Such willingness to do manual work was inherited by her youngest daughter, and was essential to Marie's success many years later in isolating radium." The intensity with which the Curies stare at us from their public portraits suggests that what Einstein said of Marie might have applied to Pierre as well: "Madame Curie is very intelligent but has the soul of a herring, which means she is poor when it comes to the art of joy and pain."

The two cold fish had sought each other out from among a flashier school of broadly cultivated, anticlerical scientists collected by the Sorbonne at *fin de siècle*. Those glittering mathematicians, physicists, and chemists formed the phalanx of the positivist movement and became a vanguard of the dreyfusards. As might be expected, the Sorbonne positivists became the targets of the protofascist Right and the fans of *La France profonde*. As part of a successful attack on Marie Curie's nomination to the *Académie*, Leon Daudet chimed in against the professors in his *l'Action française*:

> They are all like that fanatic Poincaré, a man of genius, they say, in mathematics but stupid and hateful. When it comes to the rest; the Jew of color photography Lippmann, that fanatic Dreyfusard Appell, dean of the faculty of Sciences . . . [they] no longer hide behind the life of the Saints, but behind algebra, physics and chemistry treatises. . . . They intend in fact to chase from the house all who don't think like them, don't feel like them, who have the audacity not to deny God, not to insult Rome, go to mass, to raise their children as Christians.

Nowadays they are ignored in cultural histories of the Third Republic (Jerrold Siegel's *Bohemian Paris* and Eugen Weber's *Fin-de-Siècle Paris* come to mind). But from the laws and units with which their names will forever be associated, their discoveries remain part of the fabric of twentieth century science. For me, the lives and work of these dreyfusards of science constitute a monument to reason that their contemporaries in the arts have not quite matched. Official France agrees: their names are woven into the fabric of Paris. The square before the École Polytechnique (4eme) is named after the dashing but very married Paul Langevin, whom Einstein said would have discovered relativity had he himself not done so and whose affair with widowed Marie almost cost her the second Nobel prize. His was the most noble career in the resistance—and he suffered for it. The square before the Sorbonne (4eme) is named after mathematician Paul Painlevé, who was Langevin's second at the duel he fought to preserve Madame Curie's honor. Painlevé was another early supporter of Einstein; he was politic enough to become a minister of war. Langevin quipped that Painlevé had studied Einstein thoroughly, unfortunately not until after he had written about him, a sequence acquired perhaps in politics. Marie's other partisans in the Curie–Langevin scandal dot the landscape as well: Gabriel Lippmann, the Nobel physicist who invented color reproduction and who presented the Curies' discovery of radium to the *Académie des Sciences*, is remembered behind the Place de la Nation (20eme); Paul Appell, dean of the faculty of sciences, has an Avenue of his own near the Cité Universitaire (14eme)—it leads to the Avenue Rockefeller; the Rue Henri Poincaré loops off Blvd. Gambetta (20eme) and Emil Borel is off the Blvd. Periphérique in the 17eme.

The story of Pierre and Marie Curie is a tribute to a dazzling set of discoveries jointly made by a man and a woman of genius. *"L'Art c'est moi; la science c'est nous,"* wrote Claude Bernard, and—feminist pleaders aside— that *nous* remains independent of gender, psychic baggage, or family romance. Among the most memorable photographs in Quinn's book is a late one of an intense Marie Curie on the balcony of her Institute behind the École Normale. Her lined face looks forward to the future, her hands are scrumbled by the scars of radium; it's an image that sums up the hope and the harm of her discovery. She would have been pleased that she is shown overlooking the street that we now call the Rue Pierre et Marie Curie.

8

THE WOODS HOLE CANTATA

Each August, when pink mallows fill wetlands by the bay, the Woods Hole Cantata Consort gives its annual performance. In recent years, chorus and orchestra have ventured the Bach "Magnificat," Haydn's "Creation," Handel's "Alexander's Feast," and tonight, the "Gloria" of Vivaldi. The concert is given in the fieldstone Church of the Messiah, which presides over a snug churchyard almost at the verge of Vineyard Sound. Inside, the church has the trim, no-nonsense bearing of its nautical setting: bright timber work, well-hewn pews, and high brass. On the festive nights of these concerts it becomes the intersection of at least two cultures.

The performers and their audience are drawn chiefly from the scientific summer community of the Marine Biological Laboratory, with help from year-round residents and the occasional semi-professional soloist. Led by Elizabeth Davis, wife of Harvard's eminent microbial physiologist, the performers run in age from adolescence to retirement; many of their surnames are familiar to readers of *The Journal of Biological Chemistry, The New England Journal of Medicine,* and *The Proceedings of the National Academy of Sciences.* Twenty minutes before the performance, the audience is already tightly packed into the pews; children and limber postdocs squat in the aisles. Latecomers will have to sit on the brick steps or the green lawn to hear the music through open doors. The audience—relatives, friends, coworkers, students—fills the hall with the tribal buzz and chatter that one hears at class reunions or graduations. The air is laced with pizzicati of nervous laughter that I recognize from my children's first recitals at music school or their undergraduate theatricals. As the performers file in, I look about at the community gathered here. One can identify embryologists whose winter habitats range from Hawaii to Naples, biochemists from

Northwestern to Stony Brook, physiologists from Seattle to the Cambridges, physicians from Duarte to Lund. Tomorrow, in the laboratory down the hill, some of their number will trail the flow of ions through the squid's giant axon or impale the eye of a crab; others will watch granules explode in the egg of a sea urchin. Some, I imagine, may simply look up from a journal in the library and stare, in sheer puzzlement, at the tossing sea. Tonight, though, we are all here *en famille* to celebrate the ancient ritual of music in concert.

A hush before the opening bars; then, in a rush of sound, we are surrounded by the throb of an amateur chorus in full summer voice:

Gloria in excelsis Deo!

This splendid noise and its reverberations move us along for a while. I follow the text, agreeing in principle with the general sentiment:

Et in terra pax hominibus
bonae voluntatis

The music churns gloriously on. Frail in attention, my thoughts move from melody to text—and beyond. It occurs to me how unique an occasion this evening presents—what a series of clashing interests have been, literally, harmoniously resolved. The "Gloria" is Catholic liturgy, the church Episcopalian. The composer was a priest, the majority of performers are probably freethinkers. Vivaldi is Art, the audience lives Science. Under the roof of one church sit Moslem with Jew, Indian with Pakistani, Harvard with Yale: a truce is obtained, the conditions for which seem to have eluded a good bit of mankind. Nowhere have I seen it recorded—to turn to the clash of the Two Cultures—that late August evenings in the Hamptons, in Big Sur, or Woodstock are devoted to responsive readings from Darwin's *Origin of Species* or Watson's *Molecular Biology of the Gene*. Nor, to return to clerical considerations, have I heard that religious enthusiasts have sat in rapt attention as, in joyous phalanx, they chant sections of Diderot's *Rameau's Nephew*, H. L. Mencken's essays, or Jacques Loeb's *The Mechanistic Conception of Life*.

The chorus modulates to

Domine Deus, Rex coelestis
Deus Pater Omnipotens

and my wayward thoughts turn, almost irretrievably, to Jacques Loeb, whose spirit remains so alive in this community. I had, indeed, almost tripped over him on my way to the concert; his simple gravestone—marked only with name and dates (1859–1924)—lies in the churchyard just a few feet from the rear entrance. What irony that the ashes of Loeb should come to rest by the Church of the Messiah! For Loeb, according to his biographer, R. L. Duffus, was champion of the antispiritual, a freethinker

> who believed that "living organisms are machines and that their reac- tions can only be explained according to the same principles which are used by the physicists"; who declared that nature was a blind muddler, with whom "disharmonies and faulty attempts are the rule, the har- monically developed system the rare exception"; who taught that consciousness and free will are illusions; who reduced the lord of creation to the status of a chemical solution bubbling in a tube; and who recognized nowhere in the universe a purpose or a God.

Yet there in the churchyard were his remains, cheek by jowl with the more God-fearing families of this small New England village. Loeb lies among the Swifts, the Stuarts, and the Fays, close to the founders of Woods Hole biology: F. R. Lillie, C. O. Whitman, and E. B. Wilson. A considerable distance separates his remains from those of more recent additions to this necropolis of the eminent: Selman Waksman, Hans Einstein, and Stephen Kuffler. We owe the presence of these distinctly non-Yankee names to an enlightened Church and community, who have turned the little graveyard into an ecumenical cemetery of biology à la Père Lachaise. But Jacques Loeb in a churchyard? The supreme mechanist, Loeb was persuaded that

> life, i. e., the sum of all life phenomena, can be unequivocally explained in physico-chemical terms.

In his dazzling career as an experimental biologist at Strasbourg and Würzburg, then Bryn Mawr, Chicago, Berkeley, and the Rockefeller Insti- tute, spelled by long summers at Woods Hole, he combined imaginative science with a tough, antimetaphysical stance in matters of philosophy. He took as first principle that

> our existence is based on the play of blind forces, and [is] only a matter of chance. . . . We ourselves are only chemical mechanisms. . . .

Loeb presented this view (printed as the first chapter of *The Mechanistic Conception of Life*) at the high-water mark of pre-World War I scientific optimism, the First International Congress of Monists at Hamburg in September of 1911. The meeting, a veritable "concourse of the omniscients," according to Donald Fleming, attracted "to this, the most radical, libertarian, and anti-Prussian of German cities, delegations of freethinkers, freemasons, ethical culturists, socialists, pacifists, and internationalists." And to this audience of several thousand, Jacques Loeb outlined the truths of the new quantitative biology, of which he was so much a part. The work of Mendel and Morgan had established the mathematical rules of heredity, Perrin and Millikan had experimentally shown the *existence* of atoms and molecules, and Loeb himself had shown not only that the instincts of lower animals could be redefined as "tropisms," but also that one could approach the challenge of creating life in the lab by parthenogenesis. What need for a sentient *Pater omnipotens*, when, as Loeb predicted in his address, future scientists will doubtlessly achieve "the task of producing mutations by physico-chemical means . . . and prove that they have succeeded in producing nuclear material which acts as a ferment for its own synthesis and thus reproduces itself."[*]

With biology firmly established on the quantitative terms of physics and chemistry, with behavior a predictable response to chemical tropisms, with the beginnings of artificial life sputtering at the bottom of a beaker, all that was left to explain was the place of values in this world of fact. Loeb asked:

> How can there be an ethics for us? The answer is, that our instincts are the root of our ethics and that the instincts are just as hereditary as is the form of our body. We eat, drink and reproduce not because mankind has reached an agreement that this is desirable, but because, machine-like, we are compelled to do so. . . . We struggle for justice and truth since we are instinctively compelled to see our fellow beings happy.

Loeb was convinced that both individual and group behavior followed from genetically programmed "instincts" as surely as the markings on the

---

[*]Loeb's prophecies were fulfilled when Hermann J. Müller induced mutations by X-rays in 1926; the blueprints of the nuclear material and its reproduction were drafted by Francis Crick and James Watson in 1953, and the ferment (DNA polymerase) was described by Arthur Kornberg in 1955. These discoveries were rewarded in the most appropriate manner, at Stockholm.

wings of fruit flies followed from the rules of Mendelian genetics. And genetics would soon have a chemical basis. Evil was simply a mutation or an uncommon error:

> Economic, social and political conditions or ignorance and superstition may warp and inhibit the inherited instincts and thus create a civilization with a low development of ethics. Individual mutants may arise in which one or other desirable instinct is lost, just as individual mutants without pigment may arise in animals; and the offsprings of such mutants may, if numerous enough, lower the ethical status of a community. . . . Not only is the mechanistic conception of life compatible with ethics: it seems the only conception of life which can lead to an understanding of the source of ethics.

It is difficult not to be charmed by this generous summary. Loeb here suggests not only that ethics derive from genetics but also that an inheritable instinct to make our fellow men happy guarantees the "struggle for justice and truth." Believing that a genetic blueprint specifies the virtue of our biological machine, Loeb is persuaded that "low development of ethics" results from rare mutations or unusual "economic, social and political conditions." This Panglossian view of the gene machine is, of course, as dated as the scientific optimism of the beginning of this century. The brutal wars and senseless murders which followed so closely upon the hopeful assembly at Hamburg have given sufficient evidence, I should have thought, that the direct opposite of Loeb's hypothesis is equally likely to be true. Verdun and Coventry, Dachau and Nagasaki make it just as likely that the blueprints of "instinct" specify brutish conduct and that the ethical drive for "justice and truth," which seemed so natural to Loeb, may constitute the aberrant mutation.

Loeb saw himself as heir to a rich patrimony of rational thought dating from the Enlightenment. Son of Benedict Loeb, a well-off Jewish importer who had settled in the Rhineland, he was raised in a secular, cultivated home of which the major intellectual heroes were the humanists of eighteenth-century France. Throughout his life, Jacques Loeb looked to the philosophes as his guide; indeed, his *The Organism as a Whole* (1916) was dedicated to Denis Diderot. And when the whirlwind struck, when the ship of scientific optimism was sunk by the guns of August, he remained true to the spirit of

that group of freethinkers, including Alembert, Diderot, Holbach and
Voltaire, who first dared to follow the consequences of a mechanistic
science, incomplete as it then was, to the rules of human conduct, and
who thereby laid the foundations of that spirit of tolerance, justice and
gentleness which was the hope of our civilization until it was buried
under the wave of homicidal emotion which swept through the world
in 1914.

I look about in the church as the chorus turns to the *Miserere*. A list of
young Americans killed in that wave of homicidal emotion hangs on the
wall. The music rises:

Qui tollis peccata mundi
miserere nobis

Loeb's public responses to the war constitute some of the more lucid
formulations of liberal opposition to slaughter on the Western front. He said
that the war made him "sick," and framed an emotional plea against the
jingoism of the time in "Biology and War," which was published in *Science*
on January 26, 1917.

The biology of which the war enthusiasts make use is essentially
antiquated, and so we need not be surprised to find that they consider
war to be based on what they call the "biological law of nature," or the
"survival of the fittest."

The war enthusiasts also derive from what they are pleased to call the
"law of nature" the statement that "superior races" have the right of
impressing their civilization upon "inferior races." The information
concerning the relative values of races is furnished by a group of writers
who call themselves "racial biologists." This "racial biology" is based
on quotations from the erudite statements of theologians, philologists,
historians, politicians, anthropologists, and also occasionally of biologists, especially of the nonexperimental type . . . the sad fact remains
that this pseudobiology has had at least a share in the production of the
tragedy which is being enacted in Europe. For wars are impossible
unless the masses are aroused to a state of emotionalism and fanaticism,
and the pseudobiology of littérateurs and politicians *may serve this*
purpose in the future as it has in the past [my italics].

Contrast this view, if you will, with the majority sentiment, as articulated by William Osler in *"Aequanimitas"*.

Loeb was provoked not only by the senseless war, but also by the racial nonsense chucked about by "pseudobiologists" on both sides. He had left Germany in 1891, partly because of his political and social opposition to Prussian orthodoxy and partly because the *numerus clausus* made it even more difficult than in America for a Jew to scale the academic ladder. He may, therefore, have been more sensitive than many to the waves of national intolerance that rolled over his adopted country as it entered, fought, and finally won the war.

The prodigious pace of Loeb's research did not slacken in wartime. Indeed, his work—which had won him international fame and gained him the general ear—became more exacting and more quantitative. But, in addition, during the war years of 1914–1918 he took up his pen to plead before a general audience the liberal ideals of pacifism, tolerance, and understanding, and to attack the misuse of poorly perceived biology. In 1914 came "Freedom of Will and War" in the popular *New Review*; "Science and Race" in the transient *Crisis*; in 1915 came "Mechanistic Science and Metaphysical Romance" in the *Yale Review*; in 1916, *The Organism as a Whole: From a Physico-Chemical Viewpoint*; in 1917 he published "Biology and War."

After 1917, however, there was only public silence from Loeb. Had he simply stopped trying? Were reason and scientific optimism insufficient weapons with which to fight the accelerating mischief? Why had the new Enlightenment not bettered the "economic, social and political conditions, or ignorance and superstition" which had dragged the most advanced countries of the West into self-destruction? One can sense from a letter to his friend, the great Swedish chemist Arrhenius, some of the despair Loeb felt in the xenophobic climate of postwar America:

> Politically I think America is in a bad way. We are suffering from a wave of reaction and from religious and racial fanaticism just as the European countries do . . . I wish something might be done to make the world a little more promising for the next years, but the outlook is bad. Personally I work hard because I *want to forget* and I am very grateful to science that it permits us to forget a good deal of the outside world. [my italics]

It seems likely from this, and other letters, that the lessons of war changed Loeb in a profound way; and I think that I can document the consequences of that change by a little bookkeeping from his bibliography. The last year before the war, 1913, was a productive one for Loeb; it was the year that he and Reinhard Beutner correctly deduced that the potential

electrical difference which exists across biological membranes might be
maintained by phospholipid layers. In 1913, he published sixteen papers,
and of these, fourteen described experiments with living animals. In 1920,
an equally productive year, Loeb published eighteen papers but only two
described research on living creatures. "I want to forget," wrote Loeb, and
left his prickly sea urchins to launch studies on purified proteins *in vitro*.
These important essays into colloidal chemistry were to occupy him until
his death.

This drive to reductionism, which has characterized the biological
revolution of our time, fits (ironically) Loeb's description of a "tropism."
Tropisms have long been discredited as adequate explanations for the wide
range of human behavior; they cannot, for example, explain our deepest
psychic lives, our art, or our music. But to me, they seem sufficient to
describe the more limited behavior of scientists at their job.

Loeb first encountered tropisms in 1888; their analysis established his
reputation. Certain caterpillars emerge in the spring and move to the tips
of branches in order to feed on the forming buds. This behavior, before Loeb,
had been generally attributed to the playing out of an innate, almost
metaphysical, drive for self-preservation. Loeb showed that this "instinct"
had a distinct biological basis in photosensitivity. Since the perception of
light is, by definition, a chemical event—a photoelectric cell can tell if a light
is on or off—instincts are nothing but chemical reactions. Loeb found that
if the only source of light with which he provided the caterpillars was from
a direction opposite that of the food, the caterpillars would move to the
direction of the light—and starve. Loeb interpreted these data on heliotro-
pism to show that free will could be ruled out as motive for *any* action; all
behavior had roots that could be reduced to biochemical responses. I view
Loeb's experiment as the perfect description of what scientists do in a
laboratory; they crawl in heliotropic wriggles to the source of light. Often
that light shines from the wrong direction, but follow it they do, and with
results anticipated by caterpillars.

Quoniam tu solus sanctus
tu solus Dominus

The Vivaldi is soaring to resolution. The church, warmer than at the
beginning of the evening by virtue of summer, of an enthusiastic audience,
and of modest ventilation, is filled with ravishing chords of the Baroque. I
become persuaded by the thought that Loeb was wise, indeed, when he

decided to forget a good deal of the outside world and to devote the last six
or so years of his life to establishing the precise isoelectric point of gelatin.
Admired by the freethinkers of his time (Thorstein Veblen, Sinclair Lewis,
H. L. Mencken), Loeb and his generation of mechanists survive only in the
memories of their students. Unfortunately Paul deKruif was one of these,
and deKruif's popular accounts of Loeb made the mechanistic conception
of life sound like a sophomoric attempt at village atheism. Mechanism did
not outlast Prohibition.

But it is possible to argue that mechanistic philosophy, which drove
Loeb and his generation into transient fallacies of social thought, has
produced some of the noblest and most lasting products of our civilization.
By acting "as if" the descriptions of man and nature can be reduced to the
blueprint of machines, our generation has broken the genetic code and
landed on the moon. As Loeb would have predicted, by deciding that not
only our actions, but also our thoughts, are governed by rules (structures)
which control mechanical or electronic machinery, we have transformed the
continents into matrixes of microchips and the social sciences into schools
of structural engineering. Loeb would have been surprised that this "tropism" to physical and chemical models of the natural world has probably
failed to advance the sum total of human happiness over that in 1913. We
now know, or at least believe, that the simple, rational, anticlerical positions
of Diderot and the philosophes, or of Loeb and the mechanists, are no more
likely to resolve the problems of war and violence than is the text of the
"Gloria." And while mechanistic philosophy may describe adequately how
science works, it does not offer consolation for the world it produces:
Vivaldi may be more appropriate to that task. Nevertheless, whether we
like it or not, all experimental scientists are mechanists now. Ravished by
Art in this church tonight, we will wake in the morning to work in the mills
of Fact, the construction of which we owe to the mechanistic conception of
life.

Si monumentum quaeris, circumspice ["If you seek his monument, look
about you"] reads a tablet at the entrance to the main building of the Marine
Biological Laboratory. The quotation, directed at the work of Christopher
Wren, comes from St. Paul's Cathedral. The tablet is dedicated to F. R. Lillie,
who lies near Loeb in the churchyard. The building is named "Lillie," and
it is at right angles to one called "Loeb," where dozens of students are
enrolled in the modern version of the physiology course that Loeb taught
in the 1890s. Loeb's own tablet, near Lillie's, reads:

JACQUES LOEB
1859–1924

BRAIN PHYSIOLOGY
TROPISMS, REGENERATION
ANTAGONISTIC SALT ACTIONS
ARTIFICIAL PARTHENOGENESIS
COLLOIDAL BEHAVIOR

I am pleased that this community of science acts each day as if Diderot and Loeb were right, that we look at our jobs as the solution of one mechanical problem after another. I am also reassured that we are all here listening to Vivaldi, whose final notes cannot, by any stretch of the imagination, be ascribed to tropisms. Perhaps that is why the ashes of Loeb lie buried outside this monument to the nonmechanistic, the irreducible: this church filled with music and the students of Loeb's students. After years of teaching, of writing, of reasoning, the shade of Jacques Loeb may have earned the right to listen in quiet satisfaction to this harmony.

But tomorrow, when oscilloscopes flash in the lab, when the electrodes twitch with the signals of a nerve, when life precipitates from white threads of DNA, Loeb's monument will rise about us. The flesh may perhaps be weak, but the spirit of mechanistic science survives as long as one postdoc dips a sea urchin egg into a beaker of salt. What rest by the Church of the Messiah are the ashes of Loeb, the disappointed optimist. His monument is the mechanistic conception of life, the manifesto of a biological revolution which has spread far from the cozy waters of Woods Hole.

Cum sancto spiritu in gloria Dei Patris.
Amen.

9

FOUCAULT AND THE BAG LADY

On cold Mondays in February the victims of winter cluster at Bellevue. Infarction and pneumonia, exposure and gangrene drive the inhabitants of street and park to its doors. It was therefore a routine event when police brought the woman we'll call Mrs. Kahaner to the emergency room. Suffering from apparent frostbite, she had been found at the side entrance of Lord & Taylor early on Sunday morning and was admitted with a temperature of 93 degrees Fahrenheit.

The nurse's notes described her belongings. She was evidently the prototypical bag lady. Her rags, pots, and jumbled impedimenta overflowed five shopping bags crammed into a wire cart. An unkept appointment slip found in her purse identified her as a former patient in one of the New York State mental hospitals from which she had been last discharged three years ago. A telephone call to that hospital yielded her provisional diagnosis (chronic schizophrenia), her medications on discharge (the usual phenothiazine derivatives), and her age (fifty-six). The nurse had guessed Mrs. Kahaner to be an unkempt forty-five or so. The telltale areas of black gangrene on her fingertips had almost directed her admission to the surgical service, but a more thorough examination by an astute medical resident brought her to the medical wards, where our Rheumatology Service was called to see her the next morning.

The resident had observed that the skin over Mrs. Kahaner's face was drawn so tightly that she seemed to be peering out from behind a mask of fine leather. Over her chest and extremities, patches of pearl white skin alternated with areas of café-au-lait pigmentation. Her digital gangrene was only an extreme sign of impaired circulation—even after she had been warmed to normal temperature, her hands and feet remained mottled purple and pink. Her hands were also deformed by pincerlike contractures, her mouth was dry, and her eyelids crusted. X-rays showed fibrosis of the lungs, an enlarged right side of the heart with a prominent pulmonary artery segment, and diffuse calcification of the soft tissues. The admitting diagnosis was scleroderma.

As we examined her, Mrs. Kahaner remained resolutely silent. Warmed, hydrated, with oxygen aboard, she slowly responded to those requests necessary for physical diagnosis, but not at all to questions of medical or personal history. Frightened and withdrawn, this thin woman with sparse, stringy hair looked at us as if ready to weep, but tears did not come. As we were quick to appreciate, her dry eyes were a consequence of the disease: the early engorgement and later exhaustion of salivary and lacrimal glands known as Sjögren's syndrome, which arises in the course of scleroderma. Indeed, so hidebound was the skin about her lids, that the lacrimal glands could not be expressed. And her relatively youthful appearance was due to the shiny reflection of her collagenous mask, which had lost its normal wrinkles and fissures. We fitted the known stigmata of her disease to the patient before us. It turned out that she displayed the characteristic involvement of skin, lungs, and blood vessels that results when normal connective tissue—a pliant web of complex sugars, tensile proteins, and clear capillaries—is replaced by a carapace of scar tissue: scleroderma.

Despite extensive research, we know next to nothing about the cause and care of this sometimes progressively fatal malady. We are unable to stop its major assaults on vital organs. And although we have learned to manage the crises of breathing, vessel spasm, and kidney failure that mark its progression, the underlying problem of scleroderma is as unsolved today as it was thirty years ago, when I first learned about the disease in medical school. Monographs have been published, hundreds of research reports have flowed into our journals, and clinical subsets and related syndromes have been described. But while we are now able to discuss with greater precision the effects of scleroderma on the major organs, we remain as helpless as the doctors who forty years ago watched a network of collagen fibers tighten inexorably about the hands of Paul Klee.

As the weeks progressed, our patient responded to symptomatic treatment: vasodilators, artificial tears, hydration, and a soft, palatable diet. A caring staff managed to enter into a kind of communication with Mrs. Kahaner. This enterprise was made difficult by her persistent belief that she was bedded in the municipal hospital of Reykjavik, and that the Icelandic police were monitoring her words and behavior. This delusional system and associated fantasies were elicited by the house staff and consulting psychiatrists, who with patience and a dram or two of the newer psychotropic drugs dampened her fears of the Arctic constabulary. She was put at vague mental ease for the first time in months.

Nevertheless, the further management of her psychiatric troubles eclipsed her somatic complaints when the time came to plan her disposition after discharge. It was clear to the medical staff that she required close, periodic monitoring of her scleroderma, and equally clear that she belonged in some sort of institution that would shelter and clothe her, and where her madness would be managed in such a fashion that she could avoid the injuries of weather and a violent city. She needed what she had never found, an asylum—in the original sense of that word: sanctuary, from the Greek word *asylon*, meaning "inviolable."

However ignorant our profession may be of the proper management of scleroderma, it is safe to say that we know less of the disorders we call schizophrenia. Mrs. Kahaner, it turned out, had been in and out of various mental hospitals for at least two decades; she was divorced, and no relatives could be traced. Her recent confinements had lasted only a month or two; her prompt discharges were attributed to "remissions" induced (so her keepers believed) by various combinations of a broadening pharmacopoeia. After each discharge, she appeared with less regularity at follow-up clinics. For the past three years, interrupted only by two admissions to the emergency room of Roosevelt Hospital because she had been mugged, she divided her nights between the doorways of department stores and the rest rooms of Pennsylvania Station.

Mrs. Kahaner had joined that subclass of "deinstitutionalized" mental patients who have congregated in our cities in a kind of behavioral mockery of the consumer society. These shopping-bag ladies follow trails of private acquisition not too dissimilar from those of their saner, middle-class sisters. From the avenues of the West Side to the doorway of Brooks Brothers, from the arcades of the subways to the entrance of B. Altman's, the sad battalions of mad ladies course our streets, loaded with possessions tucked in tattered paper bags. And the bags still carry their message of fashionable competition: *Bloomingdale's! Bonwit Teller! Bergdorf Goodman!* The bags, in anarchic disorder, are in turn stashed in that other symbol of our abundant life: a shopping cart from the supermarket. There is probably no other country in which the madmen and madwomen so neatly exhibit the claims of local enterprise. I cannot recall similar public displays of private goods among the cat ladies of Rome, the mendicants of Madrid, the down-and-out of London and Paris. Madness is surely as prevalent in those cities as in New York, but mockery of The Shopper in the person of the bag lady strikes me as pure Americana.

However, the disposition of poor Mrs. Kahaner after Bellevue will be governed by fashions in attitudes toward the mentally ill that know no borders. Motivated by the observation that the overt aberrancy of the badly disturbed can be managed by the wonders of psychopharmacology, our asylums have evolved over the past decades into homes of only temporary detainment. The therapeutic rescue fantasy—as current in Paris as in Glasgow, in London as in Albany—has been joined to the concept that the diagnostically insane will be more humanely treated, or achieve greater personal integration, if they are permitted freely to mix with the "community."

It has always seemed to me to constitute a fantastic notion that the social landscape of our large cities bears any direct relationship to that kind of stable, nurturing community which would support the fragile psyche of the mentally ill. Cast into an environment limited by the welfare hotel or park bench, lacking adequate outpatient services, prey to climatic extremes and urban criminals, the deinstitutionalized patients wind up as conscripts in an army of the homeless. Indeed, only this winter was the city of New York forced to open temporary shelters in church basements, armories, and lodging houses for thousands of half-frozen street dwellers. A psychiatrist of my acquaintance has summarized the experience of a generation in treating the mentally deranged: "In the nineteen-fifties, the mad people were warehoused in heated public hospitals with occasional access to trained professionals. In the sixties and seventies, they were released into the community and permitted to wander the streets without access to psychiatric care. In the eighties, we have made progress, however. When the mentally ill become too cold to wander the streets, we can warehouse them in heated church basements without supervision."

Clearly, what is lacking is the concept of asylum: a space where the mad and deranged can be protected not only from their internal demons but from the harsher brutalities of the street. Our loss of the asylum as an ideal can be traced to a number of social vectors, many of which take origin in the most altruistic of motives. Deinstitutionalization—a horrid word—became practical when the advent of palliative chemotherapy made possible the release of inmates whose overt behavior, on discharge, posed no immediate danger to patient or citizen. But despite the obvious fiscal and administrative relief afforded the state and local authorities, release would not have become widespread were it not for the vigorous efforts of well-meaning advocates of civil rights. These activists were rightfully convinced that commitment of the unwilling to mental institutions might represent the first step toward a Soviet-style incarceration for dissidents. Moreover, the return of the mad to the community was in accord not only with the rosy visions of fellowship and liberation that became prevalent in the sixties but also with a general distrust of arrangements made by any authority, be it juridical, governmental, or medical. And a prominent role was played in the attack upon the concept of asylum by radical sociologists, psychiatrists, and historians.

Absent any real knowledge of the nature of schizophrenia, one attractive path out of the forest of ignorance would be to deny that the term itself had meaning. It was not surprising, therefore, that the revisionist psychiatrist R. D. Laing seized the temper of the sixties and argued that those called "schizophrenic" were not sick in the medical sense. To Laing and his adherents, the schizophrenic—a soul more sensitive than the ordinary person—has made an entirely appropriate adjustment to an insane world. Since mental "disease" is only definable in relationship to a fractured, social framework of bourgeois values, *any* accommodation is acceptable, and the job of the healer is to adjust the discomfited to his level of accommodation. The Laingians argued that since a century of scientific inquiry had failed to prove that madness can be adequately described by the medical model, it was time to abandon the concept of schizophrenia as a disease. And if we can dispense with the therapeutic model, why not abandon those institutions that exist to treat this now nonexistent malady?

Most physicians, unlike other "mental health professionals," cannot be easily persuaded by this argument. For one reason, the argument suffers from what has been called the reification fallacy. This fallacy mistakes the *name* of a condition, or syndrome, for its *effects*. Were we to apply such flawed reasoning to the example of scleroderma, for instance, we would have to abandon the practice of making that diagnosis in Mrs. Kahaner's case. But by so doing, one would not alter in one sad, clinical detail the narrative of how our patient became trapped by collagen. Nor would her treatment, I daresay, be vastly improved by deinstitutionalizing Mrs.

Kahaner before we had dilated her blood vessels, given her intravenous fluids, and helped her to breathe. In the example of scleroderma—a disease as enmeshed in mystery and as difficult to treat as schizophrenia—we are forced to resort to palliation, to support, and to clearheaded observation. I fail to see how our ignorance of the etiology of an illness gives us the right to assume that it is due to social maladjustment.

Indeed, the medical model may prove more effective. This model has already paid off in the case of lithium-sensitive, manic-depressive psychosis and in Alzheimer's disease. Until their basis in organic malfunction was identified, these conditions—like schizophrenia—were often attributed to social or familial wounds.

But the Laingian assault on the medical model and the modern response to that model—the asylum—has also been joined by the more urgent attack of historians and sociologists. They argue that mental hospitals subjugate and victimize the mad in the name of therapy, when what the mentally deranged require is liberation from the shame associated with the label of madness. Mental hospitals, so goes this charge, have since the seventeenth century served chiefly as schools for the brutal induction of shame; their abolition can only benefit the mad. This argument, in its most persuasive and scholarly form, is presented by the late Michel Foucault, the iconoclastic French historian of sexuality, prisons, and medicine.

In his major opus, *Madness and Civilization*, Foucault traces the history of mental institutions to the "Great Confinement" of the classical age. A royal edict of April 1656 led to the establishment of the Hôpital General, a series of Parisian hospitals for the confinement of mendicants, fools, and other idle folk of the street. These institutions, some of which were founded as centers for the sequestration of lepers in medieval times (lazar houses), became depositories for beggars, the unemployable, and the mentally feeble. In keeping with the new commercial spirit of the age, the Hôpital General and its sister institutions of England and Germany set the inmates to productive work and small manufacture in order to prevent, in the words of the royal edict, "mendicancy and idleness as sources of disorder."

Foucault describes how the confined populace replaced the leper as an excommunicated class:

> The asylum was substituted for the lazar house—the old rites of excommunication were revived, but in the world of production and commerce. It is in this space invented by a society which had derived an ethical transcendence from the law of work that madness would appear

and soon expand until it had annexed them [sic]—the nineteenth century would consent, would even insist that to the mad and to them alone be transferred these lands on which, a hundred and fifty years before, men had sought to pen the poor, the vagabond, the unemployed.

In these former lazar houses, the Age of Reason confined the socially undesirable: thief and demented, idler and aged, fool and madman. And so that strange mixture of the socially unwanted remained in chains or at forced labor until the advent of the French Revolution. It was at a branch of the Hôpital General, the Bicêtre, that Philippe Pinel struck these chains in an episode that signals the birth of liberal philanthropy. Pinel was confronted by an official of the Revolution, Georges Couthon, seeking suspects for trial. Pinel led him to the cells of the most seriously disturbed, where Citizen Couthon's attempts at interrogation were greeted with disjointed insults and loud obscenities.

It was useless to prolong the interviews. Couthon turned to Pinel and asked: "Now, Citizen, are you mad yourself to seek to unchain such beasts?" Pinel replied: "Citizen, I am convinced that these madmen are so intractable only because they have been deprived of air and liberty!"

"Well, do as you like with them, but I fear you may become victim of your own presumption." The official left, and Pinel removed the chains. The rest of the tale is the history of the modern mental hospital, at least until deinstitutionalization.

Now, one would have thought that Foucault would come out cheering for Pinel. But no. In keeping with a tradition of radical criticism that includes the names of Herbert Marcuse and Ivan Illich, Foucault reserves his hardest blows for Pinel and the therapeutic reforms of bourgeois liberalism. Pinel stands accused by Foucault of substituting the psychological chains of the medical model for the iron shackles of the old lazar houses. The madman had become trapped in the birth of the asylum, and this new, therapeutic community confronted him with a new menace in the history of madness: guilt.

> The asylum no longer punished the madman's guilt, it is true, but it did more, it organized it for the man of reason as an awareness of the Other, a therapeutic intervention in the madman's existence.

Foucault, displaying that curious affinity of advanced French thought for the punitive tableaux of the Marquis de Sade, goes on to grieve for the unchained patients of the nineteenth century. He poses a very fashionable paradox:

> The dungeons, the chains, the continual spectacle, the sarcasms were, to the sufferer in his delirium, the very element of his liberation. . . . But the chains that fell, the indifference and silence of all those around him confined him in the limited use of an empty liberty. . . . Henceforth, more genuinely confined than he could have been in a dungeon and chains, a prisoner of nothing but himself, the sufferer was caught in relation to himself that was of the order of transgression, and in a non-relation to others that was of the same order of shame. . . . Delivered from his chains, he is now chained, by silence, to transgression and to shame.

Foucault's analysis of the relationship between keeper and inmate, between authority and the governed, extends to the interaction between doctor and patient. He points out that life in the asylum constituted a microcosm in which the values of a bourgeois state were symbolized: relationships between family and child, centering on authority; between transgression and punishment, centering on immediate justice; and between madness and disorder, centering on the social contract. Since the doctor stood *in loco parentis* for the whole society, Foucault suggested:

> It is from these [relationships] that the physician derives the power to cure, and it is to the degree that the patient finds himself, by so many old links, already alienated in the doctor, within *le couple médecin-malade*, that the doctor has the power to cure him.

This strikes me as a remarkable passage. To begin with, one can note that the French expression *le couple médecin-malade* has the same general meaning as the English "doctor–patient relationship," but in French it acquires a sexual overtone in its reference to the marriage couple. Secondly, Foucault adds the trendy spice of "alienation" to the principles of therapy: the madman, already alienated from the old bourgeois game of power and property, recapitulates that experience with his doctor. Finally, Foucault seems to equate alienation with the concept of transference. In psychoanalysis, transference—that strong bond between patient and therapist—is a positive element in self-discovery and possible cure. The only bond identified by Foucault is that of alienation, and this gives the doctor, as locum tenens of an alienating system, the "power to cure." But for Foucault, a cure is only a partial blessing, since it implies loss of freedom. Foucault has thus shifted the therapeutic setting from the Freudian one-on-one interview to the parade ground of the asylum, where the doctor as drillmaster instructs alienated conscripts in rules of the barracks.

In all aspects of his criticism of the asylum, Foucault uses as his point of reference a golden age of madness, which for him was the medieval age, where fools and crazies formed an integral part of the unregulated life of the street. Here, Foucault's vision of integration merges with that of the antiauthoritarian Left of the sixties. Foucault and the civil libertarians view the meliorist attitudes of mental health professionals as a destructive force in the battle for self-realization. They consider prisons, asylums, hospitals, and their squadrons of social workers, psychiatrists, and psychologists as the elements of a police state designed to censor the self-expression of the mad.

In the golden age of Foucault's medieval city, fools and madmen added to the richness of everyday life by their unique insights and startling behavior. The "reforms" of the Age of Reason destroyed this organic fabric and turned it into a straitjacket. That is the charge of Foucault, heard by the intellectuals of the West as the asylums have been emptied, mental health budgets cut, and the church basements filled.

But, I should have thought, there is little chance that poor, mad people find in our society even the glint of a golden age. In the cities of America, where the Mrs. Kahaners wander outside asylums in solitary danger, where violence is unchecked, and where the aggressive roam in quest of drugs and easy victims, we have, instead, partly reverted to the Hobbesian state of nature. I am frightened by that state and frightened for the army of fools and madmen we have let loose in it. To counter this condition, I, too, have a vision of a golden age, which by no means corresponds to the present-day treatment of the mentally ill. For me, it is the dream of the Age of Reason as articulated in the liberal philanthropy of Pinel. He gave us his vision of a true asylum for those shocked by the wars of the mind, an asylum that is, perhaps, more truly civil than that which Foucault presents. Pinel, in his 1801 treatise on the nature of madness, describes the hospital at Saragossa, where there was established

> a sort of counterpoise to the mind's extravagances by the charm inspired by cultivation of the fields, by the natural instinct that leads man to sow the earth and thus to satisfy his needs by the fruits of his labors. From morning on, you can see them—leaving gaily for the various parts of a vast enclosure . . . sharing with a sort of emulation the tasks

appropriate to the seasons, cultivating wheat, vegetables, concerned in turn with the harvest, with trellises, with the vintage, with olive picking and finding in the evening, in their solitary asylum, calm and quiet sleep. The most constant experience has indicated, in this hospital, that this is the surest and most efficacious way to restore man to reason.

When we next meet Mrs. Kahaner, now lost in the shuffle between discharge from Bellevue and her unkept appointments at Creedmore, huddled in the closed doorways of Lord & Taylor, we might well ask whether her fate in our fractured city is better guided by the philanthropic vision of Pinel or the trendy critique of his dream by Foucault. Until reason and science unlock the shackles of her illness, we need, I believe, to give rest to this bag lady in a sanctuary, a therapeutic community—an asylum.

10

IN QUEST OF FLECK

SCIENCE FROM THE HOLOCAUST

In 1947, there appeared in the *Texas Reports on Biology and Medicine* (9:697–708) a report by Ludwik Fleck on "Specific Antigenic Substances in the Urine of Typhus Patients." Its opening sentences are probably unparalleled in the annals of biological science:

> The search for specific antigenic substance in the urine of typhus patients was initiated in Lwow, 1942, under German occupation. The original plan was to elaborate a test giving earlier diagnosis than the Weil-Felix reaction. . . . In addition to elaborating a diagnostic test, it was also thought to utilize the urine of typhus patients as a source of specific antigens for the preparation of a preventive vaccine, very urgently needed at this time.
>
> In May 1942, the results were reported at a staff meeting of the "ghetto" hospital in Lwow, and, several months later, the author was deprived of his collaborators who were destroyed by the Germans.

The report goes on to describe the efficacy of this vaccine in typhus fever, the second-leading cause of death in the concentration camps.

> The author, his collaborators, and 32 volunteers were vaccinated. . . . Later, 500 people in a concentration camp at Lwow were vaccinated. With few exceptions statistics of the vaccinated were unfortunately lost. Records are available only in regard to the author, his family, and two other persons, all of whom contracted typhus and recovered after a mild or abortive course of the disease. A large number of the vaccinated in the camp did not contract typhus although they were exposed to typhus infections. In contrast, the majority of the nonvaccinated prisoners contracted typhus with a fatality rate of 30%.

Now, neither the field of study described in this report (the serology of typhus) nor the journal in which it is reported would ordinarily engage my attention. However, the sequence of events that led to my encounter with this unusual medical scientist seems worth recounting, because it reflects the tides of discovery and rediscovery, the charting of which, ironically, constitutes Fleck's chief legacy to the history of science.

The development by Ludwik Fleck of a practical vaccine against louse-borne typhus from the urine of the afflicted was directly responsible for his survival in the camps. The Germans, as anxious to protect their troops as they were indifferent to the fate of their captives, forced Fleck first to produce this vaccine in the ghetto hospital of Lwow (formerly called Lemberg) and later in the camp "hospitals" of Auschwitz and Buchenwald. But in the course of his studies of the serological response to his vaccine, he made a signal scientific discovery, one that directly illuminates my own field of research, which is the role of white blood cells in inflammation. Using as his tools nothing but the clinical microscope, a few common laboratory dyes, and the (unfortunately) abundant peripheral blood of patients with typhus, he described the phenomenon he called "leukergy."

The first description of leukergy also appears in the same issue of *Texas Reports* in an article by Fleck and Z. Murczynska (the issue was devoted to reports of medical research carried out in wartime Poland). The authors present evidence that the white blood cells (neutrophils) in the blood of patients with typhus clumped into tight little clusters after a few minutes of incubation at 37 degrees Centigrade. Fleck went on to determine that these white cell clumps were also observed in the blood of rabbits that had been experimentally infected with certain bacteria or their endotoxins and, indeed, in the course of many infectious diseases. This response of the white cells, he concluded, prepared them for such other functions as sticking to the walls of small blood vessels, squiggling out of the vessels toward offending microbes, and engulfment of the bugs.

The description and analysis of leukergy, first encountered in the unspeakable setting of the death camps, occupied Fleck and his collaborators throughout the postwar period. Publications resulting from this work appeared not only in the Polish literature and in translated form in the *Texas Reports*, but also in more widely read journals, such as the *Schweizerische Medizinische Wochenschrift* and *Acta Haematologica* (Basel), in *Le Sang* (Paris), and in the *Archives of Pathology* (Chicago). In 1949, Fleck was awarded the scientific prize of the city of Lublin for his discovery of leukergy, which he classified as an important biological response of white blood cells to injury or infection. It is therefore remarkable that this work had no impact whatsoever upon the scientific community at large. Indeed, a search of the *Science Citation Index* (Philadelphia) for the years 1965 to 1980 reveals fewer than half a dozen references to this discovery.

But leukergy is very much alive, although living under assumed names. In the past decade, studies from many laboratories, including my own, have shown that the "aggregation" and the "adherence" of neutrophils are among their very early responses to a variety of inflammatory insults. The study of the function of white blood cells in disease is now an enterprise that must engage several thousand investigators throughout the world. This group, or "thought collective" (to use Fleck's phrase), has come to appreciate that the altered surface properties of neutrophils, their aggregation in blood vessels—as reflected by their clumping in the dish—probably accounts for some of the more lurid complications of allergy, infection, and shock. Hundreds of articles, scores of learned reviews, and extensive book chapters document this role—all without mention of Fleck. The trial vessel of leukergy has sunk without a trace.

If this be the fate of Fleck's major experimental opus—his work on typhus vaccine has not only been superseded but rendered superfluous by advances in hygiene, new pesticides, and the advent of antibiotics—how did the phenomenon of leukergy emerge from the depths of obscurity? The answer lies in the revival of Fleck in another context: as a philosopher and historian of science, as a forerunner of Thomas Kuhn, and as an important thinker in the sociology of science. In 1935, Fleck had published a book entitled *Entstehung und Entwicklung einer wissenschaftlichen Tatsache: Einführung in die Lehre vom Denkstil und Denkkollektiv* (*Genesis and Development of a Scientific Fact: Introduction to the Study of Thought Style and the Thought Collective*). The book, written by a Jew, was unpublishable in Germany and therefore was published in Switzerland. Only 640 copies were printed: 200 were sold, and of these one found its way into the library of Harvard. While still a member of the Harvard Society of Fellows, the illustrious philosopher and historian of science, Thomas Kuhn, came across Fleck's book, stimulated by a footnote in a work of Hans Reichenbach. Kuhn, who refers to Fleck in his *Structure of Scientific Revolutions*, recollects: "In twenty-six years I have encountered only two people who had read the book independently of my intervention."

It was at a meeting of another Society of Fellows, that of New York University, of which Kuhn was also a transient member, that I first heard mention of Fleck the philosopher. Consequently, when a new edition of Fleck's book appeared, with a foreword by Kuhn, well edited and annotated by Thaddeus J. Trenn and Robert K. Merton, I was eager to grapple with it. The book was challenging on three levels. First, Fleck presents an original analysis of how scientific facts, especially those of biology, are necessarily contingent upon the social context in which they are established. A discovery is both the product of, and a factor in, a discrete social setting. Second, a brief biographical sketch provided by the editors describes the harrowing circumstances under which the author's major discoveries were made: a tale of triumph plucked from horror. And finally, a footnote in the book, which refers to Fleck's work on leukergy, rings the bell of recognition for someone who has spent the better part of his scientific career worrying about white cells. The book deserves our first attention.

Genesis and Development of a Scientific Fact argues that scientific "facts" are not absolute, but relative, and that they necessarily relate not only to the general social scene but especially to the "thought styles," or modes of perception of those individuals who compose the "thought collective" in a particular field of inquiry. Fleck points out that the so-called Vienna Circle of epistemologists, such as Moritz Schlick and Rudolf Carnap, who were trained in the physical sciences, considered that thought processes (cognition) were *fixed* and absolute, a view not unlike that of the new structuralists. On the other hand, they considered empirical facts to be *relative*. In contrast, the philosophers with a humanistic, or sociological, background (Emile Durkheim, Lucien Levy-Bruhl) considered scientific facts to be fixed, whereas human *thought* was relative. Fleck asks:

> Would it not be possible to manage entirely without something fixed? Both thinking and facts are changeable, if only because changes in thinking manifest themselves in changed facts. Conversely, fundamentally new facts can be discovered only through new thinking.

Fleck went on to suggest that cognition itself is essentially a social activity, since the existing stock of knowledge exceeds the range available to any one individual. Knowledge is not generated *in vacuo* by a particular consciousness, or by any one person, but by a thought collective, which he defines as a community of persons mutually exchanging ideas or maintaining intellectual interactions. . . . It also provides the special "carrier" for the historical development of any field of thought, as well as for the given stock of knowledge and level of culture (thought style).

In a telling analogy, he recapitulates this theme:

> If an individual may be compared to a soccer player and the thought collective to the soccer team trained for cooperation, then cognition would be the progress of the game. Can an adequate report of this progress be made by examining the individual kicks one by one? The whole game would lose its meaning completely.

This mixture of epistemology and social theory is stirred by means of a detailed analysis of a scientific "fact." Fleck, a trained serologist, analyzed the history of the Wassermann reaction, the first blood test for syphilis. This test was his "fact," and he traced its roots to earlier notions of impure blood. Fleck compared what August von Wassermann and Carl Bruck wrote about their discovery years after their first observation with what they had originally described.

In their first two papers on the subject, Wassermann and coworkers had described use of a serological method (the complement fixation reaction) to identify *antigen* in watery extracts of syphilitic tissues. Antigens, of course, were supposed to signal the presence of the disease-causing microorganism. Positive results were obtained in sixty-four of seventy-six extracts. An *antibody*, that which is detected by the test as now performed, was found in only forty-nine of 257 samples of blood from syphilitics. Moreover, it was soon determined that the *antigen*, later identified as a ubiquitous fatty material called *cardiolipin*, was not only found in diseased, but also present in healthy, tissues. However, thanks to refinements in the technique introduced by Julius Citron and others (they used alcohol or acetone to make antigen), the test for *antibody* became positive in up to 90 percent of syphilitics. This is the "fact" of the Wassermann reaction.

Fleck quotes Wassermann's hindsight fifteen years after the fact: "I proceeded from the idea, and with the clear intention of finding a diagnostically usable amboceptor (antibody)." And after comparing this statement with the original work, Fleck concludes:

> The ultimate outcome of this research thus differed considerably from that intended. But after 15 years an identification between results and

intentions had taken place in Wassermann's thinking. . . . From false
assumptions and irreproducible initial experiments an important discovery has resulted after many errors and detours. The principal actors
in the drama cannot tell us how it happened, for they rationalize and
idealize the development.

Recent accounts of such major discoveries as the helical structure of
DNA or the discovery of the hepatitis-associated antigen would tend to
confirm Fleck. But how were these "false assumptions and irreproducible
initial experiments" transformed into an "important discovery"? Fleck is
clear:

> It was the prevailing social attitude that created the more concentrated
> thought collective which, through continuous cooperation and mutual
> interaction among the members, achieved the collective experience and
> the perfection of the reaction in communal anonymity.
>   Laboratory practice alone readily explains why alcohol and later
> acetone should have been tried besides water for extract preparation,
> and why healthy organs should have been used besides syphilitic ones.
> Many workers carried out these experiments almost simultaneously,
> but the actual authorship is due to the collective, the practice of coop-
> eration and teamwork.

Further description of the richness of this book would require much
more extensive documentation. Suffice it to say that Fleck gives the best
account, by far, of how the products of bench science become translated into
journal articles and then integrated into textbooks. Only in texts, removed
from the realities of quotidian work, do "facts" exist as such—and even
these textbook facts have finite half-lives.

I would, however, like to dwell on one passage which deals with the
main problem inherent in Fleck's view of facts as servants of collective
fashion. What is the role of the original observer, the adventurer, whose
sudden insights anticipate the consensus of the thought collective? Here we
may find the judgment of Fleck, the sociologist, on Fleck, the philosopher,
and Fleck, the biologist:

> Such scientific exploits can prevail only if they have a seminal effect by
> being performed at a time when the social conditions are right. . . . Had
> Vesalius lived in the twelfth or thirteenth century he would have made
> no impact. . . . The futility of work that is isolated from the spirit of the
> age is shown strikingly in the case of . . . Leonardo da Vinci, who
> nevertheless left no positive scientific achievement behind.

In one sense, the rediscovery of Fleck by the philosophers and historians—and what I am certain will turn out to be his rediscovery by the thought collective of the neutrophil world (at least if one reader has anything to do with it!)—is the validation of this passage and of Fleck's major thesis. The appropriate thought collectives within the social sciences and human pathology were not ready to incorporate Fleck's own contributions, not until Kuhn had brought about his own revolution, not until Merton had turned the glass of social analysis to the life of the laboratory, and not until we had learned how infections and endotoxins activate the neutrophil.

Although brief, Fleck's biography, as outlined in the Trenn–Merton volume, evokes completely the agony of a Jewish intellectual who was torn between Polish and Germanic cultural traditions, finally to be betrayed by both. Born in 1896 in Lemberg, he attended Austrian secondary schools. He received his medical degree from the University of Lwow in 1922, shortly after the town reverted to Poland. His postdoctoral training in serology was with Rudolf Weigl of Lwow and, later, at the University of Vienna, where Clemens von Pirquet and Friedrich Kraus were major figures. In 1928, he became head of a government bacteriology laboratory in Lwow and appeared to have achieved security, until he was dismissed (in 1935) in consequence of one of the not uncommon anti-Semitic seizures of the Polish people. Between 1935 and 1939, he was forced to earn his living by establishing a private, diagnostic laboratory of microbiology. Throughout the thirties his research interests (which had always been directed toward typhus, the Wassermann reaction, and host defense mechanisms) became joined to a broader interest in the social and humanistic features of the scientific calling, and he published several papers on history and philosophy, both before and after his 1935 book.

When Stalin and Hitler divided Poland, Lwow became Russian, and perhaps due to his progressive (or at least anti-Fascist) views, he was promptly made director of the city's microbiological laboratory, finally gaining an appointment to the medical school. When the Germans occupied Lwow in 1941, he withdrew, as prescribed, to the ghetto hospital, from where his first observations on both the typhus vaccine and leukergy originated. In 1942, having involuntarily ceded the methodology for producing typhus vaccine to the Germans, he was sent to Auschwitz, where his sisters and their families perished. In 1944, Fleck was transferred to Buchenwald—still in the camp "hospitals," still precipitating antigen from urine to immunize against typhus. He was liberated by the American army in 1945 and, on his return, was received with honor by the Polish authorities. From 1945 on, he rose in the ranks of the new Marie Sklodowska Curie University of Lublin to full professor; in 1952 he became Director of the Department of Microbiology and Immunology at the State Institute in Warsaw. Despite many honors, including election to his country's Academy of Science in 1954, he had always wished to emigrate to Israel, permission for which was granted in 1957. He was appointed head of the section of Experimental Pathology of the Israel Institute for Biological Research and appeared headed toward a new burst of scientific and belletristic activity when he was stricken with Hodgkin's disease. He died in June of 1961.

I cannot say with certainty what the ultimate role of this medical amateur will prove to be among the lions of the social sciences. I am certain, however, that in my own field, the rediscovery of Fleck will prove useful. He correctly perceived that the sticking of neutrophils to each other— brought about, for example, by bacterial toxins—is the first step in *Activierung des Leukocytären Apparates* (activation of the leukocytic apparatus). Activation of this apparatus permits the stimulated cell to increase its locomotion, its engulfment of bacteria, and its power to kill the microbes. It will not be difficult, utilizing the technical tools and nomenclature fashionable in our modern thought collective, to repeat, extend, and incorporate the observations made by Fleck and his coworkers on the blood of patients with typhus.

Perhaps Fleck would have been pleased that leukergy has again been perceived as a "fact." (The image comes to mind of the thought collective carrying their triumphant captain off the soccer field.) But the final irony is this: If Fleck is correct in his book, then his discoveries of the typhus antigen and of leukergy were, of necessity, born in the social crucible of the Holocaust. The bacteriologist of the ghetto hospital of Lwow would probably have been the first to wish that his discoveries were unnecessary.

11

AUDEN AND THE
LIPOSOME

Visitors to my office at Bellevue Hospital occasionally identify, with surprise, a photograph of W. H. Auden hung amidst the usual diplomas, family pictures, and group shots of house-staff days. The poet is shown next to Erika Mann; their pairing is a minor document of the troubled thirties. The marriage, in 1935, of Thomas Mann's daughter to Auden, a British subject, conferred a happier nationality upon the stateless refugee from Hitler's Germany. In the picture, Auden wears a rumpled lounge suit; his wide lapels and gaping jacket sleeves are signals of the period. His left hand dangles the perpetual cigarette, the right hand drapes an open jacket to display the braces and skirted trousers of a time when:

> *National Service had not been suggested*
> *O-Level and A were called Certs*
> *Our waistcoats were cut double-breasted*
> *Our flannel trousers like skirts.*

> from "A Toast"

The poet's hair is neatly trimmed, the face is unlined—no hint appears of those Icelandic crags and furrows that in later years were to transform his features into a relief map of the anxious age. His wife-in-name-only appears to be generating a conventional smile for the photographer. She is dressed in a fashionable windowpane frock, topped by a gay, polka-dotted coat. Her scarf is knotted in chic display, the hat is soft and rakishly tilted. The overall effect is of upper academe. Indeed, the young English master and his wife have been snapped by a photographer from the school newspaper of a progressive private school.

A. D. Bangham (photographer), *W. H. Auden and Erika Mann at the Down School,* 1935.

There are two reasons why this picture hangs above my desk: the subject and the photographer. Auden's photo reminds me that what I do for a living—medical research—may begin as fun but has a social bite. And since the snapshot was taken by my sometime collaborator, Alec Bangham, I am reminded of a long Cambridge summer in the sixties when he first taught me to form liposomes and which I remember chiefly as an interlude of pure joy.

Auden himself was persuaded that science, like poetry, is a "gratuitous, not a utile, act, something one does not because one must, but because it is fun." However, it is not this aesthetic approach to the doing of science—the approach of a skirt-trousered amateur—that has engaged me in Auden's poetry and prose. I think, rather, that I respond to that mixture of appreciation and fear of modern science which informs so many of his fabrications. He is at once a lyric enthusiast of our profession—a flatterer of the enterprise—and a necessary critic of our social mischief. His oldest friends—Christopher Isherwood, Cyril Connolly—have called him a schoolboy scientist at heart; Stephen Spender acclaimed him as the diagnostician of our fears. Son of a physician, familiar with the winners of glittering prizes (his phrase) from the laboratories of Oxbridge and the New World, he paid even the least distinguished of scientists an extravagant compliment that is difficult to forget:

> The true men of action in our time, those who transform the world, are not the politicians and the statesmen, but the scientists. When I find myself in the company of scientists, I feel like a shabby curate who has strayed by mistake into a drawing room full of dukes.

Given the political convictions of his youth, so different from his predecessors—Yeats, Eliot, Pound—this generous appraisal should come as no surprise. Auden and his fellow anti-Fascists of the thirties were convinced that the journals of science contained clues to the fellowship of man. Auden believed that the laws of physics govern servant and master alike, and that it was the job of the poet to instruct both in the language of their common history.

> As biological organisms made of matter, we are subject to the laws of physics and biology: as conscious persons who create our own history we are free to decide what that history shall be. Without science, we should have no notion of equality: without art no notion of liberty.

With these attitudes in tow, Auden devoted much of his energy to warning us of the wretched use to which both poetry and science had been put in our time, in decades during which

> The night was full of wrong,
> Earthquakes and executions,
> And still all over Europe stood the horrible nurses
> Itching to boil their children

from "Voltaire at Ferney"

Auden was ashamed by the extent to which the children of art and science enlisted in the service of brutality, injustice, and moral squalor. Commissioned as a major at the close of the Second World War, he visited in the course of his work the concentration camps where the methods of science were mocked. He spent long evenings in Bavaria recapitulating

> The grand apocalyptic dream
> In which the persecutors scream
> As on the evil Aryan lives
> Descends the night of the long knives . . .

from "New Year's Letter 1939"

In the suburbs of Munich—in Dachau—Professors Pfannenstiel of Marburg, Jarisch of Innsbruck, and Linger of Munich had frozen scores of inmates to death and reported carefully detailed autopsies to "proper" scientific congresses. Here too, Professor Beiglbock of Berlin forced Poles and Jews to drink an excess of seawater: descriptions of the victims' hallucinations and heart failures were exactly recorded in what passed for scientific manuscripts. At the Natzweiler camp, Professor Dr. Eugen Haagen—formerly of the Rockefeller Institute—worked to transmit viral hepatitis from prisoner to prisoner and managed successfully to kill several hundreds with experimental typhus. These examples of scientific disgrace were paralleled in the realm of the arts by the complicities of Heidegger, the gangs of Bayreuth and Oberammergau, by the films of Leni Riefenstahl. After such excesses, Auden became persuaded that our best hope lay in the establishment of limits, limits to the collaboration between intellect and the tyrant, best expressed in his "Ode to Terminus," the Roman God of Limits:

> In this world our colossal immodesty
> has plundered and poisoned, it is possible

> You still might save us, who by now have
> learned this: that scientists, to be truthful,
> must remind us to take all they say as a
> tall story, that abhorred in the Heav'ns are all
> self-proclaimed poets who, to wow an
> audience, utter some resonant lie.

This plea for limits seems appropriate to our new era of biological engineering and belletristic extravagance. But I sense, perhaps, another strain here: a restatement of "Without science, no equality." For Auden is speaking to us from the experience of a generation which had used the discoveries of physiology and biochemistry as a kind of shield against the biological determinism of the old Fascists. He correctly discerned in the ongoing genetic arguments based upon insect behavior (first ethology, now sociobiology) a nasty trend toward the spinning of tall tales of inequality:

> Bestiaries are out, now
> Research has demonstrated how
> They actually behave, they strike us
> As being horribly unlike us.
>
> Though some believe (some even plan
> To do it) that from Urban Man
> By advertising, plus the aid
> Of drugs, an insect might be made.
>
> No, Who can learn to love his neighbor
> From neuters whose one love is labor
> To rid his government of knaves
> From commonwealths controlled by slaves?

> from "Bestiaries Are Out"

These verses anticipate my own misgivings about recent attempts to offer the stunning successes of modern biology—our ability to decipher the social code of bees and the genetic code of man—as excuses for undoing the notion of equality. We have detailed the biochemical errors which cause blacks to suffer from sickle cell anemia, Italians and Greeks from thalassemia, Jews from Tay-Sachs disease, and Nordics from pernicious anemia; such heritable flaws speak of biological inequality. Since the popular geneticists of race and behavior have become persuaded that social characteristics also reside in the genes, is it any wonder that neo-Fascists, such as Alan de

Benoist of France, have seized upon the recent hypotheses of sociobiology and "selfish genes" to legitimize their political fantasies? Auden's worries *matter*: the tall stories of our most recent science are beginning to have a dirty fallout. Private discoveries have public consequences. That homily brings me to the second reason for the picture on my wall, which evokes not only a summer of fun but yet another worry that Auden did not live to articulate.

As I've said, Auden and Erika Mann were photographed by A. D. Bangham, F. R. S., of Cambridge. In 1935, Alec Bangham was photographer for his school newspaper at the Downs School and had been assigned to photograph the English master and his new bride. The photo remained imprinted upon a glass negative, stored among the juvenalia of this gifted amateur photographer, until Alec produced it at a scientific conference in the English countryside held fifteen years after the discovery of liposomes. It was at this conference that half a dozen investigators agreed that the use of liposomes in the treatment of human disease was not only desirable, but imminent. Liposomes are small fatty vesicles, made in the laboratory from off-the-shelf chemicals, which Alec and his collaborators originally proposed as models of cell membranes. They were soon found to duplicate many of the properties of the natural bilayers of lipid which enclose the ferments and nucleic acids of living cells. Since liposomes are biodegradable and not at all toxic, it has been suggested that they might function as the long-searched-for vectors by means of which entrapped substances might be safely delivered to organs deep in the body. It took no great effort of the imagination on the part of the conferees at this liposome meeting to tell each other tall stories of the use of liposomes for the manipulation not only of disease, but of the genes. While such experiments are only on the drawing board right now, it may not be premature to put a second concern on the agenda of angst.

I am afraid that in the decade and a half since liposomes were first constructed, developments in biological engineering have come so far, and so fast, that we are on our way not only to explaining, but to perturbing, the fundamental properties of living things. In this decade, when schoolboys and stockbrokers know how to assemble genes in the lab, we have learned to worry not only about the political consequences of our theories, but about the biological sequelae of our experiments. We have launched on an endeavor which may eventually realize the prophecy of Diderot:

> If anyone wants to describe . . . the steps in the production of man or
> animal, he will need to make use of nothing but physical agencies . . .
> eat, digest, distill in a closed vessel, and you have the whole art of
> making a man.

Indeed, the distilling of lipids in a closed vessel is a fair description of
how liposomes are fashioned. We begin by dissolving lipids, fatty materials
identical to those which our own cells use to fabricate their membranes, in
chloroform, and then dropping this solution into a closed vessel. The
chloroform is evaporated off and the fats remain as a turbid, dry film at the
bottom of the vessel. Next comes an operation which approaches magic,
and which, each time I perform it, carries with it faint intimations of the
Book of Genesis. We add a watery solution which contains any one, or
several, of the purified large molecules of life: enzymes, hormones, genes.
Then, in obedience to the laws of physics and chemistry, the fats spontane-
ously enclose these molecules in membranes of predictable geometric array.
The suspension assumes an opalescent sheen as the membranes—part
liquid, part crystal—swell with their cargo. By simple separative procedures we can then isolate liposome-entrapped materials from those which
have escaped capture. Eventually we can hold in our hand—or at least in
the collection flask—lipid-entrapped enzymes or nucleic acids: things arranged very much as they would be in a jumbled, rudimentary cell or
organelle.

I've described the formation of these little vectors of enzyme or gene in
detail because they constitute one example of how the playthings of the lab
have suddenly become capable of arousing not only aesthetic joy, but moral
qualms as well. Indeed, our colleagues of molecular biology, who can now
stitch genes in the dish and harness bacterial energies for the production of
human proteins, are now engaged in efforts at introducing their genetic
artifacts into the cells of mouse and man. Some of them are probably already
toying with the use of liposomes as vectors for bioengineering. I suppose
that when I express moral qualms at this possible application of our
discovery, I'm only saying that neither Alec nor his collaborators signed
up for this sort of activity when liposomes were first made—when one
first enthused about the prospect of actually replicating a fundamental
unit of life, a membrane capable of entrapping the stuff of cells. Now, there
is every reason to believe that in the decades to come, when second- or
third-generation liposomes can be appropriately designed to pass safely
through body fluids and deliver their contents to vital organs or tumors,

their use will prove of benefit in the treatment of disease. But every increment in our capacity for fiddling with the nature of things should, I believe, make us pause for moments of serious self-doubt, should make us worry that we are not only engaged in mischievous tinkering.

The current mistrust of scientific research stems, in my view, from four major insults to our moral sensibilities. I've already alluded to two of these: the unfounded confusion of modern genetics with social Darwinism, and the not unrelated abuse by German doctors of human experimentation. To these may be added two others: the consequences to our offspring of the "poisonings and plunder" of the earth (radioisotopes and chemical pollutants), and now, fears as to the restructuring of man by the well-intentioned splicing of his genes. These accusations are not unfounded, and as a community, we in the sciences should be prepared to acknowledge our share of the guilt, without being hobbled by the admission.

It may be scientific hubris to worry about the small contribution of liposomes to the game of genetic roulette—this will probably be played whether or not our stake is critical to the transport of inheritance. The current prospects for changing our natural load of disease and aging still seem somewhat dim—and we cannot be certain that strategies based on gene splicing or liposomal delivery are even headed in the proper direction. The enterprise is only *about* to be launched. But although the enterprise itself merits concern, I do not believe that we should back off. When I was first in Cambridge, I was present at a fastidious discussion between E. M. Forster and some young transatlantic visitors on his essay entitled "Two Cheers for Democracy." (The mostly radical visitors seem to have been concerned that Forster was giving one cheer too many.) Well, at this point, I'd like to sound two cheers for biology!

It is certainly *possible* that errors and disaster will accompany our attempts to alter the biology of man: but that biology includes diabetes, childhood leukemia, crippling arthritis, and inexorable senescence. Our technical triumphs may change the matter of the natural world, but that world maintains pandemic influenza, endemic parasites of gut and liver, and the natural carcinogens of plant and virus. In the days before the early bioengineers of microbiology (Pasteur, Ehrlich, Koch, and Metchnikoff) began to manipulate the fundamental nature of an ecosystem composed of man and microbes, the natural world contained smallpox, diphtheria, poliomyelitis, tuberculosis, and cholera.

A socially prescient humanist of 1880 might well have worried about the future of man in a world freed of microbes by the microbe hunters. He would have warned us of overpopulation, of an aging populace, of consequent famine, inflation, and social unrest. While acknowledging, in partial guilt, that some of these consequences flowed from the discovery of antibiotics, would we have been wiser *not* to conquer infectious diseases? My answer, since I am now alive thanks to antibiotics, has to be negative. Like it or not, on an actuarial basis, you, dear reader, and I are alive thanks to a society that permits the risk of error and invention, that encourages private inquiry to be expressed as public gain: the whole shooting match of Western invention and activism. Those lucky enough to be supported on the playing fields of science should worry hard—before, during, and after the game— but the rules of our sport, of science, are not written in the language of our guilt. The language which we need to remind us of that guilt remains the language of the artist, the poet, the philosopher, and—ultimately—the citizen. Auden has suggested that the language of science and that of poetry are at opposite poles; we neglect the latter at our own peril, because:

> Scientific knowledge is not reciprocal like artistic knowledge: what the scientist knows cannot know him.

So I suppose that I am reassured by that picture on my wall, as I see Auden and Erika Mann looking at the photographer, the young Bangham, and through him at me. As Auden's image in the photograph overlooks my laboratory impedimenta: the many journals, monographs, and reprints, I am persuaded that he knew what scientists are up to, that he knew the extent to which we are guilty, expressing this in language as clear as the genetic code:

> *This passion of our kind*
> *For the process of finding out*
> *Is a fact one can hardly doubt*
> *But I would rejoice in it more*
> *If I knew more clearly what*
> *We wanted the knowledge for,*
> *Felt certain still that the mind*
> *Is free to know or not.*
>
> from "After Reading A Child's Guide to Modern Physics"

NO IDEAS BUT
IN THINGS

February is a bad month for doctors in and around New York. Colds hit the young, pneumonia collects the old. Exposure saps the drunk and the homeless. Hospital beds are full and telephone lines are busy; it's dark when we awake and darker still when we get home. It's tough to get around the streets when they're clogged with snow, the roads become caked with brown slush. But no matter how bad it may seem nowadays, it must have been worse a generation ago before the new vaccines and antibiotics. Patients were less willing—or able—to come to the office or clinic and doctors spent more time making house calls, a good bit of that time served behind grimy windshields. If you want a whiff of that sort of winter—of that sort of life—try William Carlos Williams of Rutherford, New Jersey. Robert Coles, the eloquent Harvard psychiatrist, has just brought Williams' medical fictions back into print. But don't stop there. Try his *Autobiography*, or the epic five books of *Paterson*. If you're hooked by then you'll want to revisit his many other volumes of poetry, his novels, essays—you may, in fact, become as preoccupied with this physician/poet as I have been lately. Which brings me back to winter, the winter of 1948, as described by his biographer, Paul Mariani.

The "meadows" of New Jersey, ruined flatlands and industrial suburbs which lie between the Passaic and Hudson rivers, had been covered by snow since the blizzards of early December. At one-thirty on the morning of February 10, Dr. Williams, almost sixty-five, climbed into his Buick, which was parked in the hospital lot. Lot and roadside were banked by yard-high walls of snow, and as he rolled toward home, he playfully took a swipe at the snowbanks with his fenders. The car became stuck in the bank and no amount of rocking would free it. No help was available, so he trudged back to the hospital and borrowed the only tool available, a coal shovel with a broken handle. Setting to work at his usual brisk pace and short temper, he became so furious in the process that he tore his shin badly by kicking the iced snowbank in anger. Hard, mindless work. Williams had stayed late at the hospital with his patients—he'd been codirector of pediatrics since 1931—and coped throughout the day with the chaff of hospital politics. He was president of the Board of Directors of the hospital and representative to it of the Medical Board.

As the doctor shoveled, there was much on his mind. In the past few months, his two careers had come to a turning point. He would soon have to retire from his hospital position. And now that his son, William Eric Williams, was about to finish his pediatric residency at New York Hospital before taking over the bulk of his father's practice, it looked as if William Carlos would finally get a chance to devote more time to his writing. For years, Williams had placed himself on a tough treadmill, indeed. Up early, he'd soon made morning house calls—by no means limited to children—in the ethnic slums which bordered suburban Rutherford. On to the hospital, where he worked till noon. Home for lunch with his wife, Floss: a small nap, then afternoon office hours. More house calls, a delivery or two, dinner, and evening office hours. The sign on his house read:

Office Hours 1 to 2—7 to 8:30
Sundays by Appointment

He wrote poems and prose in the evenings and between patients; sometimes he pulled his car to the curb and jotted down pieces in the little notebook he kept as school physician. He made almost no money at his writing, had his works printed in small editions—which he sometimes supported by income from his practice—and had not yet found a major publisher. Although by 1948 he had achieved a good bit of success among the avant-garde, he was dismissed by most influential critics. Edmund Wilson ranked him with Maxwell Bodenheim as an inconsequential figure and Williams was disappointed that his "friend" Conrad Aiken left him out of a major anthology of American poets. Perhaps now that his son was about to pitch in he could begin to write that long, important work which would gain him the reputation he felt he deserved.

Not that Williams was unknown in the world of arts and letters. Far from it: among his earliest friends were the painters Charles Demuth, Charles Sheeler, and Marsden Hartley. It is no accident that Williams came of age with modern American painting. He maintained a steady correspondence with Marianne Moore, Wallace Stevens, and Kenneth Burke. In Paris he had been photographed by Man Ray, dined with Brancusi, and had circumcised Hemingway's son. He had edited a small magazine with Nathanael West. Ford Madox Ford had founded the Friends of William Carlos Williams Society. Marcel Duchamp had played on the lawn in Rutherford. And within the past few months, Williams had gone over the proofs of *Paterson 2* with the young Robert Lowell and worried with him over the next two sections of that book. He had visited his lifelong friend and political opposite, Ezra Pound, at St. Elizabeth's Hospital in Washington, D.C., where Pound had been incarcerated for madness rather than tried for treason. But his bohemian friends in the arts knew very little of that other life across the Hudson, the life of the solid, hardworking practitioner, who drained abscesses, scraped tonsils, and adjusted baby formulas at two dollars a visit.

Eventually, by this sixty-fifth year of his life, the strain was beginning to tell. He had written to his friend Fred Miller:

> I don't quite understand why I feel so pressed, there seems to be, on the surface, nothing more than I have always handled somehow in the past, more or less successfully, but these days I'm going about in circles.

As he continued to shovel, he felt the kind of severe anterior chest pain that any doctor recognizes as either angina or infarction. He rested against the car, experiencing, perhaps in his own body, the angst of Old Doc Rivers, hero of his best short story:

> Frightened, under stress, the heart beats faster, the blood is driven to the extremities of the nerves, floods the centers of action and a man feels in a flame. . . . That awful fever of work which we feel especially in the United States—he had it. A trembling in the arms and thighs, a tightness of the neck and in the head above the eyes—fast breath, vague pains in the muscles and in the feet.

Despite the continued pain, Williams kept digging. Finally, the car began to budge and he was able to get into it. Perhaps only another doctor can imagine what went on that night. Surely, Williams must have known that every textbook of clinical medicine contained a description of the classic candidate for angina or infarction: a sixty-five-year-old male who shovels his car out of the snow in February. And since the pain did not leave

him as he drove back to Rutherford he must have been aware that he had
suffered a heart attack. But home he drove, not back to the hospital! Home
to Florence Herman Williams, his wife since 1912, home to 9 Ridge Road,
where his mother Elena had ruled a world of spirits in her invalid room for
many of her hundred years.

We have no record of what Williams thought that night, as he returned
to home and bed. Certainly the pains persisted as the Buick rolled downhill
to Rutherford. Did he expect to die, there in his car, in that mobile world in
which he spent so much time and which featured in so many of his stories?
The tale of his alter ego, Doc Rivers, begins with: "Horses . . . For a physician
everything depended upon horses. They were a factor determining his life."
Williams' own multihorsepowered Buick, now carrying his life, had seen
these roads before at all seasons and all hours. Its journeys to "Guinea Hill"
permitted Williams "entrance to the secret gardens of self"—those encounters to which he attributed the strength of his writing.

Many of *The Doctor Stories* begin at the end of a journey by an unnamed
physician/narrator to the house of a patient, where the encounter between
doctor and patient teaches the doctor—and the reader—something entirely
unexpected. Doc Rivers makes many such journeys, takes up cocaine as *his*
second career, becomes a small-town deity, and surprises us by realizing the
fantasy of every doctor who has ever been worried by house calls: ". . . he
built a fine house with a large garden, lawns and a double garage, where
he kept two cars always ready for service." In "The Girl with a Pimply
Face," the narrator over the course of several home visits to a sick baby
becomes aroused by her sister, a nymphet, ". . . Legs bared to the hip. A
powerful little animal." The baby is discovered to have a congenital lesion
of the heart, the doctor's own heart is engaged by the sister, "a tough little
nut finding her own way in the world." Finally, he is astonished to learn
from colleagues at the hospital that the girl has a "dozen wise guys on her
trail every night." In "A Night in June," the doctor drives across the tracks
to the house of an Italian mother of eight. The difficult delivery is helped
not so much by his "science"—an extract of pituitrin—but by the comfort-
ing, and expert, hands of the patient. He discovers that:

> The woman in her present condition would have seemed repulsive to
> me ten years ago—now, poor soul, I see her to be as clean as a cow that
> calves. . . . It was I who was being comforted and soothed.

In "Danse Pseudomacabre," our doctor is roused from his bed at three o'clock in the morning by a fat Scot with an infection of the face. The narrator learns that he is needed not only to minister to the patient, but to witness the will! In "The Use of Force," the doctor encounters an attractive young "little heifer" of a thing who puts up a tremendous struggle to prevent her throat from being examined. The girl provokes a rage in the narrator, who is quite aware of the sexual overtones as he confesses how "I could have torn the child apart in my own fury and enjoyed it. It was a pleasure to attack her." In the course of making the diagnosis of the patient's disease—diphtheria—the doctor has acknowledged his own.

But the doctor who drove in pain through the Jersey night was not only a writer of conventional short stories in the realistic mode, he was a poet for whom invention and imagination—the discovery of the new—was a variety of religious experience. Perhaps behind the wheel of the car that night, this secular faith was of use. From *Paterson 5*:

> *We shall not get to the bottom:*
> *death is a hole*
> *in which we are all buried*
> *Gentile and Jew*
>
> *The flower dies down*
> *and rots away*
> *But there is a hole*
> *in the bottom of the bag*
>
> *It is the imagination*
> *which cannot he fathomed*
> *It is through this hole*
> *we escape*
>
> *Through this hole*
> *at the bottom of the cavern*
> *of death, the imagination*
> *escapes intact.*

No believer in an afterlife, Williams had for over forty years put his bet on his power of invention. His goal was the forging of a new, distinctly American poetic "line." This line was to call up images of the everyday world in the patterns of real speech—what Williams called the "roar of the present"—which would separate our poetry from myth, metaphor, and allusion, from "the past above and the future below." He insisted that he was not only after images ". . . as some thought, but after line: the poetic line and our hopes for its recovery in the sense that one recovers a salt from solution by chemical action." In such an endeavor, words were to be used as pieces of type, elements of design, dispersed on the printed page like paint squirted on an unprimed canvas. To realize these complementary goals, Williams was able to draw on two cultural strains which constituted his personal history, and which framed a sort of dialectic.

In his objectivist, realistic mode, Williams paid tribute to his major predecessor at the task of uncluttering the language: Walt Whitman of Camden. But there were no lilacs in the dooryards of twentieth-century Passaic. Williams took his cues from the calls of his immigrant neighbors (*Come on! Wassa ma'? You got broken leg?*), from the cadences of the police docket (*I think he means to kill me, I don't know what to do. He comes in after midnight, I pretend to be asleep*), and from the rhetoric of American progressivism (*. . . I refuse to get excited over the cry, Communist! they use to blind us*). He was closely allied in this effort to the social realists, and especially the precisionist painters whom he knew so well: Charles Sheeler, Ben Shahn, Louis Lozowick. The work of these artists paralleled the literary efforts of the objectivists, whose press first published Williams' poems in 1934. This objectivist, descriptive strain of poetry looks as clean—as black and white— on the page as a Lozowick lithograph.

From *Paterson 1:*

> *Things, things unmentionable*
> *the sink with the waste farina in it and*
> *lumps of rancid meat, milk-bottle tops: have*
> *here a tranquility and loveliness*
> *Have here (in his thoughts)*
> *a complement tranquil and chaste.*

But Williams was not only a precise American realist, who turned things unmentionable into the stuff of tranquility. His new poetic line carried the marks of a more complex—and a more cosmopolitan—strain than any to which Williams readily confessed. He may have waged a constant war against the professional Anglophiles of English departments, he may have called T. S. Eliot's *The Waste Land* "the great catastrophe to our letters," but he was by no means an autodidact or small-town primitive. "Pop never in his life made more than the barest possible income. . . . Yet we did have an occasional case of Château Lafite in the cellar," he announced in his autobiography.

Pop was English, never acquired American citizenship, and traveled a great deal of the time in South America as a salesman of pharmaceuticals. Williams' mother was born in Mayaguez, Puerto Rico, to a French mother and a Sephardic Jewish merchant. His maternal uncle, Carlos—whose name was to give Bill Williams the touch of the poetic—was a Paris-trained physician who practiced in Port-au-Prince. William Carlos Williams was no product of the local schools of New Jersey. He had spent a childhood year at an exclusive Swiss school near Geneva, commuted across the Hudson to pre > at Manhattan's experimental Horace Mann, and made the varsity fen ing team at the University of Pennsylvania. It was at Penn that he fell in with the poets Ezra Pound and Hilda Doolittle (H.D.).

After house staff training at French and Babies' hospitals in New York, he did postgraduate work in pediatrics at Leipzig and traveled extensively in Europe. In England, Williams met Yeats in the company of Ezra Pound, who by that time had already established himself as a poetic prodigy. In Italy, he stayed with his brother Ed at the Villa Mirafiori of the American Academy, where Ed, a graduate of MIT, had won a Prix de Rome. Then back to Rutherford, where he not only plunged into practice, but into the life of the avant-garde of Greenwich Village as it caught European fire from the Armory show of 1913. There Williams was pursued through the streets by a mad German baroness, and visited salons which featured Nancy Cunard (of the ocean liners) and Vladimir Mayakovsky (of the Russian revolution). Williams spoke French and Spanish, translated from both languages, and could pass in German. All in all, this was hardly the curriculum vitae of your average suburban practitioner.

When his poetic line drew on this cultural strain of cosmopolitan energy, it carried with it echoes of "impressionism, dadaism, [and] surrealism applied to both painting and poem." The line was fractured on the page in the manner of collages, there were wild, sometimes incantory rhythms that might have derived from the lifelong spiritualism of his mother and maternal grandmother (seances in the living room!). In this continental mode, his verse ran a course parallel to the work of yet another group of painters Williams admired greatly: Francis Picabia, Juan Gris, and Pawel Tchelitchew. This piece, from *Paterson 5*, displays the two strains of Williams' muse:

Satyrs dance!
        all the deformities take wing
                Centaurs
leading to the rout of the vocables
        in the writings
of Gertrude
        Stein—but
                you cannot be
an artist
        by mere ineptitude
The dream
        is in the pursuit!
The neat figures of
        Paul Klee
                fill the canvas
but that
        is not the work
                of a child . . .
I saw love
        mounted naked on a horse
                on a swan
the tail of a fish
        the bloodthirsty conger eel.

Here is Williams doing all the things he does so well, reminding us of the work involved in setting simple words, like tesserae, on the page: this is not the work of a child! It is, in fact, the work of a mature artist who has both consciously and unconsciously grasped the two opposing strains of his nature and united them.

Williams had little use for popular psychoanalytic jargon, especially when applied to biographical or literary material. Nevertheless Robert Coles, introducing *The Doctor Stories*, recollects that Williams once compared the insights he obtained from his writings—the "descents into myself"—to the insights of the analytic experience. Riding that winter night in his car, chest aching with the fear of worse to come, was a doctor whose self-diagnoses were written in language that transcends the clinical. For Williams faced head on, in poetry and prose, the dynamics of a common enough family romance. In his autobiography he tells us of the last dream he had of his father. It is 1918, shortly after his father's death; the old man appears on the staircase of his New York office building, carrying business letters. Turning to his son, the father tells him, "You know all that poetry you're writing? Well, it's no good." Williams recalls that he awoke trembling, and in a phrase reminiscent of Dante, assures us that "I never dreamed of him since." Echoes of that bitter encounter can be heard in Williams' furious rejections of Eliot—that other Anglo-American—and eventually of Pound (who, in the wordplay of their irascible correspondence, finally became devalued).

His mother, on the other hand, played both muse and audience to his craft. In "The Artist" he describes the narrator, the poet-dancer, performing a marvelous ballet leap in the house: the wife looks in from the kitchen to ask "What's going on here?", but the mother cries "Bravo!" in surprise, and claps her hands. In life and art Williams seemed to have resolved the dialectic between the factual, English, prosaic aspects of his father's persona and the less rational, French, and artistic aspects of his mother's. He became both doctor and poet, suburban pater familias and Lothario of the studios. His poetry, too, resolved the dialectic between objectivism and surrealism, between American fact and French fancy, between patriotism and social protest. These resolutions came by way of his unique capacity for invention. And, in the manner of all great art, the poet's inventive effort leaves sweat on the brow of his reader. From *Paterson 2:*

> *Without invention nothing is well spaced*
> *unless the mind change, unless*
> *the stars are new measured, according*
> *to their relative positions, the*
> *line will not change, the necessity*
> *will not matriculate: unless there is*
> *a new mind there cannot be a new*
> *line, the old will go on*
> *repeating itself with recurring*
> *deadliness: . . .*

The invention of this sort of line—which was to be his personal achievement in verse—seems to me to have a unique, if perhaps trivial, base in his daily life. His son, Dr. William Eric Williams, made available some of his father's clinical records and notes; these have been reproduced in the William Carlos Williams Commemorative Issue of the *Journal of the Medical Society of New Jersey.* One can find there a copy of one of the typical medical charts of Passaic General Hospital, in which the case is detailed of a five-week-old baby seen by Williams in December of 1946. These sheets, on which he must have written daily, are divided by a line ruled down the middle of the chart. This compositional element, which was by no means a common feature of the medical records of the time, forces the physician/narrator to break his notes into short lines of two to three words. Thus, in the record illustrated, Williams' notes seem broken into the cadence of poetry.

The child has not done
well with its feedings
since birth

Breast feeding was
not successful

On a reasonable formula
of milk, water and D. M. # 1
a diarrhea developed
1 week ago

Ears examined
  the right found
  to be red and bulging

Sent to Dr. Wm Schwartz
to be opened.

The middle ear infection
was thought to be
the root of the problem.

Compare the final diagnosis on this chart with the final diagnosis in the poem called "Proletarian Portrait"; both follow a chain of clinical reasoning jotted down in chart-like fashion:

A big young bareheaded woman
in an apron

Her hair slicked back standing
on the street

One stockinged foot toeing
the sidewalk

Her shoe in her hand. Looking
intently into it

She pulls out the paper insole
to find the nail

That has been hurting her

In his most eloquent testament to his medical work, a chapter of the autobiography entitled "The Practice," Williams confesses what he had learned from the years of treating sick babies, their mothers, the poor, from the thousands of house calls. He discovered that the voices of these patients, these people, had given him access to the sounds of his personal poetics. They were offering the doctor, in their own voices and with their own bodies, a profound story of the self, a poem:

> The girl who comes to me breathless, staggering into my office, in her underwear a still breathing infant, asking to lock her mother out of the room, the man whose mind is gone—all of them finally say the same thing. . . . For under that language, under that language to which we have been listening all our lives a new, a more profound language, underlying *all the dialectics* [my italics] offers itself. It is what they call poetry.

Williams made it back safely to Rutherford that night, which he spent in intermittent pain. (Did he wake Floss?) Only after he rose and breakfasted, however, did his pain become unbearable. He called his cardiologist, Dr. Gold, who diagnosed a small anterior myocardial infarction. In those days, before the days of coronary care units, he was appropriately treated by six weeks of bed rest.

Relieved of medical work, Williams went on a remarkable spree of writing and publishing. He started on his autobiography, plunged deeply into Books Three and Four of *Paterson*, and, indeed, in the space of the next five years published many of the books by which we now remember him. These include the *Collected Later Poems* (1940–1950), and *Make Light Of It: The Collected Short Stories* (1950). Honors and recognition finally came his way—as if to compensate him for those long years in his thirties, forties, and fifties when Ezra Pound accused him of "pissing his life away." While still convalescing from his infarction, he was informed that he was to share the Russel Loines Award of the American Academy of Arts and Letters with Allen Tate. In 1950, he received the first National Book Award for poetry, for *Paterson 3*. And in 1953, he shared the Bollingen Prize—that prize which was first won by Pound in a sea of controversy—with Archibald MacLeish.

But the poet's life seems to have been played out to the moral tune of the old Spanish proverb "You get what you want from this world, but you pay for it!" At the peak of the recognition for which he struggled—the lectureships, the critical praise, the ardent disciples—his strength and energies were sapped by a series of strokes which affected not only body but mind. With pluck and tenacity, he learned to peck out verse with one hand, and wrote so well in the last decade allotted to him (1953–1963) that he was posthumously awarded both the Pulitzer Prize and the gold medal for poetry of the American Academy of Arts and Letters.

Williams lives. The bookshops have his works in print, the museums are full of his friends Demuth, Sheeler, Lozowick. There is a WCW newsletter. The courses of medical humanities are filled with students who read him for his subject matter alone. And in these times when androgyny is recapitulating phylogeny, many of his readers are delighted by his celebrations of heterosexual love. But the Williams who will outlast this flurry of attention is the poet of urban invention, who described the American landscape in a series of keen images which he set to a music through which we can hear the roar of the present. Unlike Pound and Eliot, he had no hankering for the past of empire and the church. Williams knew that

> —*the times are not heroic*
> *since then*
>> *but they are cleaner*
>> *and freer of disease.*

By paying attention to *things*, those bytes of clinical observation which make up the life of a people, rather than to abstract *ideas*, which in his lifetime had caused so much mischief, Williams came to terms not only with his own poetry, but with ours.

> —*Say it, no ideas but in things*—
> *nothing but the blank faces of the houses*
> *and cylindrical trees. . . .*
> *Say it! no ideas but in things. Mr.*
> *Paterson has gone away*
> *to rest and write. . . .*

NOBEL WEEK
1982

The ceremonies that crown the Nobel Prize awards in Stockholm resemble a mixture of the Academy Awards in Hollywood and a coronation. Although the names of the recipients are announced in October, the prizes are not awarded until the anniversary of the death of Alfred Nobel, December 10. A brief, handwritten will left the bulk of Nobel's estate to endow prizes in physics, chemistry, physiology, medicine, literature, and peace. The prize for peace is awarded in Oslo, also on December 10, but all of the others are given out in the Swedish capital. In the eighty-one years since the first prizes were given, the Nobel Festival (as it is referred to in Sweden) has assumed aspects of a national celebration, an international jamboree of the intellect, and a remarkable spectacle, at the core of which stands reward of human merit. It is also a great house party.

A guest, arriving early on Wednesday before the major ceremony of Friday night, is likely to be whisked to the Grand Hotel by one of the fleet of black limousines assigned to each of the laureates for a week. The grande dame of Swedish hostelry has been entirely booked by the Nobel Committee, and the front entrance is banked by the large, black Mercedes limousines, each with its identifying placard on the windshield: NOBEL-BIL. The lobby is awash in an untidy flood of radio and television newsmen. This year, the battalions are swollen by scores of Colombians; their ranking author, Gabriel García Márquez, has won the Prize in Literature. In their scurry with lenses, in their scuttle with sight lines and cameras, in their show-biz attire, which contrasts so greatly with that of the other Nobel guests, these Latin reporters appropriately remind one of Márquez's gypsies in *One Hundred Years of Solitude*.

By now it is midday. Barely time to unpack and change. Across the
water from the Grand, one can see the Royal Palace, lights already lit,
topped by blue and yellow national standards aflutter against the gray sky.
The weather is surprisingly temperate for a Scandinavian December; the
first snow has yet to fall.

By half past two, the guests of the laureates in Physiology or Medicine
are taken to the Karolinska Institutet, where these Nobel lectures will be
given. The lectures in chemistry, physics, and economics are given simultaneously before the appropriate faculties; the speech of Márquez will be
given at the Swedish Academy that evening. By two-forty-five, the main
lecture hall of the Karolinska has been packed for half an hour. Hundreds
of students, postdoctorates, and junior faculty fill the seats and spill into
the aisles. The front rows, reserved, now fill rapidly with laureates, their
families, guests, and the professorial cadre of the Karolinska: Von Euler,
Pernow, Luft, Uvnäs, Holm, Böttinger, Ernster, and others. The audience
examines itself with much craning of necks; signs of recognition and
welcome are given. For the front rows are also studded with visitors from
overseas who have been invited by virtue of their close association with
prostaglandin research or with the laureates themselves: Flower and Mon-
cada from England; Ferreira from Brazil; Griglewski from Poland; Goodman, Weisblat, and Oates from the United States; Paoletti and the Folcos
from Italy. The faces are familiar from a dozen tribal assemblies of the
prostaglandin clan.

Silence. The professor of biochemistry extends his greetings and lays
out the ground rules. Each of the laureates will give a lecture of 30 to 40
minutes, followed by a ten-minute intermission. At the close of the third
lecture, there will be a gala reception. (A glance at the audience assures the
visitor that all are suitably dressed for standing about with champagne.)
The professor introduces Sune Bergström, who is being honored for bringing prostaglandins—fatty substances which function as local hormones—
from the laboratory to the clinic. Senior of the three prizewinners, the
magisterial Bergström outlines with great modesty his achievement in
isolating prostaglandins and defining their chemical structure. As befits a
major official of the World Health Organization, he describes how pro-
stanoids can modulate fertility and how they show promise in meeting the
needs of underdeveloped countries. He addresses a fitting tribute to his
former teacher, Ulf Von Euler (Nobel Prize, 1970), now seated before him in
the center of the front row. It was Von Euler who gave prostaglandins their

name and directed the early work of Bergström. The chain of learning and scholarship is evident.

The tradition becomes yet more palpable in the room as Bengt Samuelsson assumes the podium after the intermission. For Samuelsson has not only followed *his* teacher, Bergström, in prostaglandin research but succeeded him as dean of the medical faculty. Tall, blond, and soft-spoken, Samuelsson describes the biological insight and bravura chemical analyses that, in addition to elucidating the biogenesis of prostaglandins, led to the discovery of related and equally important compounds, thromboxanes, and leukotrienes. He finishes with a dazzling summary of how these products play vital roles in the pathophysiology of circulatory disorders, inflammation, and asthma.

John Vane, of England, follows the two Swedes. With great wit and commensurate affability, the portly Vane generously splices vintage photos of his collaborators into the narrative of his main discoveries. By means of ingenious *in vitro* techniques based on those described by *his* intellectual mentor, Sir Henry Dale (Nobel Prize, 1936), Vane found that another substance related to prostaglandins, prostacyclin, was a key component in the regulation of vascular tone and the clotting of blood. He also made the heuristically important observation that aspirin and similar drugs inhibit the biosynthesis of prostaglandins. Vane finishes his lecture with a reassuring message: In an era of intricate instrumentation and recondite stereochemistry, the methods of classical pharmacology can still yield great results in the service of a prepared mind. Applause. The lectures are history.

The auditors are now shepherded through the halls of the Institutet by students dressed in folk costume. Other groups are singing. Outside, it has been dark since three-thirty, and small pots of flame outline the driveways. Inside, large candles are lit, as the passageways yield to a large, blond-wood reception hall. Hugs, kisses, or handshakes are exchanged on the receiving line formed by the laureates and their families. And then the corks pop and champagne flows. It appears that all the cellars of Rheims have been abruptly emptied for this, as indeed for all the week's receptions. As the Vikings of the Karolinska fall upon the ever-filled tulip glasses, the animated crowd mixes with little regard for the usual hierarchies of Swedish academia. Postdocs chat with laureates, teenagers with distinguished visitors, laboratory assistants mingle with professors.

The reception also serves as an introduction to the bounty of Swedish hors d'oeuvres: dried or marinated slivers of reindeer meat, coral beads of salmon caviar on sliced eggs, a most imaginative assortment of herring, and butter-soft smoked salmon. In multiple geometric disguises these appetizers will reappear in the course of receptions and dinners to follow this one. Indeed, the taste of champagne and reindeer meat will become as engraved on the gustatory memories of the Nobel guests as madeleines on Proust's. As the champagne flows and music plays, it is difficult to remember that the way to this brightly lit hall has been by means of long hours at the bench, the disappointments of false leads, and more than one hundred man-years of solitude.

By half past seven, the reception gradually breaks up, and the jet-lagged guests are dispersed to various dinners in the city. Limousines drop the elated groups into charming eateries of the Old Town snuggled in cellars behind the Royal Palace. There, too, reindeer and salmon, white wine and aquavit extend the evening.

Meanwhile, as subsequently reported in the papers, Gabriel García Márquez is addressing a packed house in the Swedish Academy, which flanks the palace. The fiery, mustachioed, and close-cropped Colombian delivers a speech charged with the energy of the Latin American left. He is responsible for the presence of Régis Debray and Mme. Mitterrand at the festival. They are in the audience as the anti-Yankee writer pays homage to his literary, if not ideological, kinsman:

> On a day like today, my master, William Faulkner, said, "I decline to accept the end of man." I would feel unworthy of standing in this place that was his, if I were not fully aware that the colossal tragedy he refused to recognize 32 years ago is now for the first time since the beginning of humanity nothing more than a simple scientific possibility.
>
> Faced with this awesome reality . . . we, the inventor of tales, who will believe anything, feel entitled to believe that it is not yet too late to engage in the creation of the opposite utopia, a new and sweeping utopia of life, where no one will be able to decide for others how they die, where love will prove true and happiness be possible and where the races condemned to 100 years of solitude will have at last and forever, a second opportunity on earth.

A few houses away, the evening is ending for the visitors from overseas, as buoyed by adrenaline and Veuve Cliquot, they toast the science of medicine. The first night ends; sleep is more than welcome.

On the next day, the first scheduled activity is a reception at the Nobel Library of the Swedish Academy at 3:30 P.M. This information is found on a personalized schedule, which is clipped to a sheaf of engraved invitations, admission tickets, and arrangements for transportation. And so, after a morning of museum hopping, the overseas visitors are brought a few blocks to the Swedish Academy. By midafternoon the sky is already darkening; a gentle rain falls. Up the stairs and past a receiving line, at which towering officials of the Academy and the Nobel Foundation offer courteous greetings, one is ushered into the Nobel Library. The room is blue and white—it is pure eighteenth century and large enough to accommodate 200 to 300 guests.

This is the first occasion at which the entire group of visiting guests and laureates has assembled. One can identify among them Ken Wilson of Cornell, the Prize winner in Physics, and the large Wilson clan. Wilson looks a bit like a solid Warren Beatty, and he is surrounded by a crowd of dazzled well-wishers. But so is his father, E. Bright Wilson, who is one of America's best-known physical chemists. And there is scholarly Aaron Klug, the Cambridge Prize winner in Chemistry, chatting in animated fashion with one of the sequined ladies of South America. Tall, bankerish George Stigler, the Laureate in Economics from the University of Chicago, chats with Daphne Vane, whose husband is busy having his hand pressed by a Nobel official. On this occasion, dress is not formal, but splendid cocktail gowns are worn by the women and sober, dark suits by the men. The ubiquitous reindeer, salmon, and champagne are dispensed in the main chamber, where under a glass case lies the original will that started the Foundation. A bust of Nobel is at the end of the room, and the guests are assured that this likeness—as well as all the other portraits and busts that dominate the interior of the Nobel Foundation—were executed from photos, since Nobel never consented to pose for painter or sculptor.

Groups of guests wander into the Nobel Library itself, where the works of Yeats nestle against those of Steinbeck, where Darwin and Dickens share an alphabetical fate. But the mix of living guests is as random as the world of learning and scholarship: Economists lift glasses with physicists; a few biochemists and some of the Colombian exiles have found a common language in French. Yesterday's reception was a tribute to the enterprise of biomedical research; today we are here to honor the general world of scholarship and literature. No speeches, no formal arrangements, simply an amiable crowd with good heads and no glitz gathered in one of the most exquisite halls this side of *Der Rosenkavalier.*

By 5:00 P.M. the reception is over, and the guests repair to a series of private parties and dinners given by the laureates or their friends in Stockholm. After the large grouping of the afternoon, the scene shifts to traditional Scandinavian hospitality. For some this proves quite grand, indeed. Thus, García Márquez dines as the guest of Olof Palme, the socialist prime minister of Sweden, with Régis Debray and Danielle Mitterrand. Also invited is a former prime minister of Turkey, Bulent Eçevit, who has been paroled by his military junta for the occasion. In contrast, other guests attend more modest receptions at the homes of the local laureates, warm gatherings of old friends and scientific collaborators. The receptions are followed by long, relaxed dinners, which end somewhat early, for tomorrow is the big day—the day of the ceremonies.

The Grand Hotel bustles with activity on the next morning. It is Friday the tenth of December. The women's hairdresser has been booked for weeks, and the scene is said to resemble the dressing room of a grand opera three minutes before the curtain is raised for *Aïda.* An American guest, to whom the staff has allotted a precious few minutes, stares with wonder as a distinguished lady, addressed as Mme. Nobel, has her coiffure studded with networks of real diamonds. The men's costumes are also being prepared. Hotel valets swish about the corridors carrying newly pressed tailcoats, white pique vests, and satin sashes as if the safety of a country depended upon their timely and unruffled delivery. In the coffee shop, Aaron Klug prepares for the evening-long ceremonies with a quick hamburger. Everyone must be ready by four in the afternoon, for the ceremonies will begin at four-thirty sharp.

The *Nobelstifelsens Hogtidsdag* (or Solemn Festival of the Nobel Foundation) is held in the Grand Auditorium of the Concert Hall. Many thousands wish to attend, but there are only 1,070 seats in the hall. The laureates are each assigned twenty-five for family and guests.

Rain is pouring at four. The lobby of the hotel is filled with guests in gala attire waiting to be transported by car or bus to the ceremonies. The pearls and diamonds, the medals and white ties are covered by cloak or fur. The busy press and television photographers light the scene with flash and spotlights. Everyone is then trundled off to the Concert Hall.

Despite the rain, a crowd of onlookers has already gathered to view the arrival of the glittering crowd. And in the lobby, a transformation is achieved. As the cloaks and capes are surrendered to wardrobe attendants, the guests are revealed as characters out of Franz Lehar: We have entered the past century. The women wear lavish ball gowns of silk and taffeta. The hairdresser's hand is evident in the ribbons and flowers braided into blond Swedish hair and in the gems that gleam from the high, white coiffures of the academic dames. There is no question, however, but that it is the men who steal the show. The printed directions for dress mandate white tie and tails. And although the Americans feel quite splendid in their rented black-and-white costumes, they are positively dowdy compared with the Europeans or South Americans. For these worthies have arrived swathed in medals and decorations: red and blue sashes worn diagonally over white vests, scarlet ribbons at the neck, enameled gold orders in the shapes of stars, crosses, and sunbursts worn over the breast or impaling a ceremonial shawl. Members of the diplomatic corps glitter like Mexican shrines; the Swedish professors doff the distinctive, academic top hats which bear not a little likeness to those of the Puritan fathers.

The scene is a gala bedlam of glitter and display; one finds it difficult to leave this scene of ritual costuming. After tiaras and sashes are adjusted, after another round of voyeuristic excess, each couple ascends a grand staircase to the main hall. The ushers are all students who, in addition to their formal dress, sport the marshal's red sash and a white student cap. The great hall fills quickly, for the ceremony begins sharply at 4:30 P.M. The stage is brightly lit for worldwide television. It is decorated with masses of flowers from San Remo, the winter refuge of Nobel. His bust is floodlit in the background. In a good-sized gallery above the stage is most of the Stockholm Philharmonic with their conductor, Sixten Ehrling. On twin banks of tiered stage seats are placed members of the Nobel Foundation and an assortment of dignitaries.

The hall is filled and the doors are now shut; the audience is hushed. Again there is a good bit of neck craning and delicate finger pointing but this is interrupted by a sound of trumpet and organ music. The royal family arrives amidst a triumphal fanfare. More trumpets, and the laureates troop in. The King and Queen, Carl Gustaf XVI and Sylvia, seem to have been brought in for the evening from either Central Casting or a fairy tale. He is tall, dark, and handsome. She is beautiful and dressed in blue and silver; ransoms of diamonds are scattered on silk and crown her hair. They are seated, stage left, on two blue chairs, which stand slightly higher than those of two other members of their party. Seated stage right in individual armchairs are the laureates, all except one dressed in white tie and tails. Márquez has chosen the permissible option of "national dress" and wears a tunic-topped, white linen costume, called a *liqui-liqui*. With his cropped hair, mustache, and tunic, he looks like the young Stalin.

More music, and the deputy chairman of the board of the Foundation rises for the first speech. He gives a stirring defense of the importance and influence of the Nobel Prizes in the world of science. He dismisses—without direct reference—a crotchety editorial in *The New York Times*, which grumbled about the winners, the donors, and the extravagance of the Nobel Prizes. Dr. Browalth replied (the remarks are in Swedish, but the audience is given a booklet with English translation): "Criticism of Alfred Nobel's donation in a recently published commentary on this year's prizes has therefore been forced to resort to the method of rewriting history, twisting and misinterpreting facts to make them fit the slanted argument. According to this opinion, the prizes are tainted because Nobel made his fortune as a maker of arms and ammunition. This is not correct. The property left by him comes from his inventions in the field of explosives, which have enabled tunnels to be driven through mountains, facilitated the construction of roads and railways, and laid the foundation of the modern mining industry." This speech is followed by Elgar's *Pomp and Circumstance*, during the course of which the audience has time to ponder the relationship between railways and artillery shells, between science and industry, between prizes and motivation, and of the distance between the Concert Hall and the place where dinner will be served.

The next speaker is Professor Lundquist of the Royal Academy of Science, who briefly summarizes the work of Kenneth G. Wilson. He then proclaims the citation: "For his theory for critical phenomena in connection with phase transitions." The laureate rises, advances to stage center, and receives his Prize in Physics (medal and scroll) from the King, who accompanies the award with a handshake. The King sits, and the laureate—alone on the stage—bows directly to the audience. No tedious words of gratitude and acknowledgment are uttered. A volley of prolonged applause. The ceremony is repeated for Aaron Klug, who receives from Bo Malmström his Citation in Chemistry: "For his development of crystallographic electron microscopy and his structural elucidation of biologically important nucleic acid–protein complexes." More music. Then Bengt Pernow outlines the work of Bergström, Samuelsson, and Vane, citing their "discoveries concerning prostaglandins and related biologically active substances." They receive their awards, one after the other, from the King and bow to massive applause. More music, Bartók this time. Literature is next: Lars Gyllesten, honoring Márquez's apparent desire to avoid the language of the Yankee in this ceremony, speaks in French to Márquez (the remarks are translated into English and Spanish) in citing the author for "his novels and short stories, in which the fantastic and the realistic are combined in a richly composed world of imagination, reflecting a continent's life and conflicts." Márquez receives his prize and, after his bow, gives a great grin, waves to the cheering audience of his adherents, clasps his hands above his head in the manner of a champion bantamweight, and resumes his seat. The music now played is by an American: William Schuman's "Chester" from his *New England Triptych*. A statement?

The last recipient is George Stigler, cited in economics for "his seminal studies of industrial structures, functioning of markets, and causes and effects of publication regulation." A bow, applause, and now the whole assembly rises as the Swedish National Anthem is played. More music; the King and Queen withdraw. The laureates remain on stage, and as the audience begins to leave, the immediate families of the prizewinners rush onto the stage to exchange embraces and kisses.

By limousine and bus, the guests are brought to the great Town Hall of Stockholm. Despite the rain, torchbearers light the road to the courtyard. They are Boy Scouts, and their wet faces shine with greeting. After entering from a cloistered *porte cochére*, the guests repeat the disrobing ritual. They are immediately given an elaborate booklet, in which each guest's name, alphabetically listed, has been assigned a number. A detailed foldout seating plan locates his assigned place at one of the tables. The banquet hall in which the *middag*, or dinner, is served is enormous: It is the height of the Metropolitan Opera in New York but larger. The style is red brick and vaguely Florentine (the Palazzo Vecchio comes to mind). Indeed, this banquet hall is still called the Blue Hall after its intended stucco color, but the architect was so impressed with the raw power of uncoated brick that it has remained *brut* ever since.

In the center of this puissant hall stands the long "A" table, set for eighty, at which will sit the laureates, the royal party, the prime minister, and various dignitaries. The other Nobel guests will be accommodated at two flanking sets of twelve tables holding thirty places each. Around this middle grouping are more than forty tables, at which anywhere from eight to forty students are placed. Their presence has traditionally been a feature of the Nobel banquet, and by this means the students are initiated into the splendors and rituals of Swedish academia.

The audience, with glitter and medals, wet hair and high spirits, find their assigned seats. One is not seated with one's spouse or even necessarily at the same table. Instructions are that both the initial conversation and the first dance request are addressed to the woman on the right. The champagne glasses are already filled, the tables are candlelit, the hall itself is lit by lanterns and reflections from day-bright television spotlights. More trumpets sound, as the royal party and the laureates, arms linked with dinner partners, descend the grand staircase from the balcony above.

After the party at the main table has been seated, toasts are proposed, first to the laureates, then to Alfred Nobel: The hall resounds with *Skoal!* General conversation swells the room, much of it directed toward the seating arrangements. Sune Bergström is next to Mme. Mitterrand, Daphne Vane is between Aaron Klug and the prime minister, Mrs. Wilson is next to the King, as is Mrs. Klug. Bengt Samuelsson is beside Señora Barcha de García Márquez. The Queen is next to John Vane; she is the target of a hundred flashbulbs. Karin Samuelsson is flanked by Ken Wilson and García Márquez. The table is filled with ambassadorial couples from the United States, Britain, Colombia, and Norway.

The guest tables are dotted with names of Sweden's great families: Nobel, Bonnier, Hammarskjöld, and so forth. The Karolinska is well represented, as are *De Aderton*—The Eighteen—who are the Swedish equivalent of the French "immortals" in scholarship and letters. There are tables for the press and dozens of Stiglers, Wilsons, Vanes, Klugs, and Samuelssons. The glasses are filled again.

Service is directed by a kind of regimental maître d', who with a wave of hands orchestrates a battalion of waiters and waitresses. The menu begins with marinated reindeer meat and progresses to a species of salmon-like trout, fresh from the Arctic, over dill and rice. The only wine is champagne: Rheims is surely dry tonight. The lights dim as a vast procession of servants descends the great staircase. Above their heads they bear the famous Nobel parfait—an elaborate concoction of two kinds of ice cream and spun sugar, each portion of which is topped by a candied N. Port is passed with dessert, and conversation rises again.

A sound of trumpets. Silence again. It is now the turn of each laureate to say a few words; the speeches are no longer than three minutes. In English, Bergström speaks of science, Vane of discovery, and Stigler of the art of economics; García Márquez speaks in Spanish of the power of poetry. Ken Wilson seizes this moment, as the United States Nobel Laureate in Physics, to call for an end to the development of nuclear technology for war. And Bengt Samuelsson brings us back to the academic basis of the pomp and glitter. He addresses the students and welcomes them to a fellowship in which scholarship and merit are their own rewards, to a fellowship of peaceful study, and to the possibility of high achievement. Applause.

A sound of trumpets. Silence, a crescendo of music, and the stairway is awash in light. An amazing procession now descends. It is led by a flag-bearing platoon of Swedish students in red sashes, tailcoats, and their distinctive white caps, but behind them comes a never-ending array of Latin American dancers, singers, and entertainers. In this tradition-breaking display, the government of Colombia, perhaps to entice García Márquez back from exile, has flown to Sweden several troupes of popular artists. These now proceed to perform what can only be described as a wild, rhythmic night club show of music, song, and dance. Drums beat in the northern hall. Colombia may have suffered one hundred years of solitude, but it cannot have endured these in silence! Tepid applause rewards these efforts.

After this unusual interlude, the customary entertainment proceeds. This is provided by a chorus of Swedish students, who sing Swedish folk songs, American spirituals, and academic anthems. Loud applause, the house lights are turned on, and the dinner is over.

It is time for the dancing to begin. The guests now ascend the great staircase that leads to the large *Gyllene Salen*—the Golden Salon. This marvelous hall is entirely faced with golden mosaic tiles, the decoration of which, by means of allegorical flora and fauna, follows a style that success-fully blends Gustav Klimt with Byzantine Ravenna. The orchestra begins with great swoops of waltzes. The floor is soon filled with billowing skirts and a flapping of tailcoats. The bars are open, the students join in the dance, and the music moves from fox trot to swing to disco as the evening unfolds.

The dances continue long after midnight; no one wants the evening to end, and guests reluctantly straggle to the waiting cars. Many roam the salons of the Grand Hotel in search of the final brandy at 2:00 A.M.

There is more on tap for the weekend. The laureates will receive their checks—$150,000 per prize—in a flutter of signatures and notaries. Later, on Saturday night, they will dine alone with the King and Queen at the Royal Palace, attended by liveried footmen. There will be small, private farewells for their overseas guests at the Grand Hotel; García Márquez will entertain his well-wishers to Latin rhythms. There will be performances by the Royal Ballet at the jewel box of an opera house and long leisurely holiday lunches at the *Operakällaren*. But one other event will touch the memories of many.

For the past decade or so, it has become customary for the Jewish community of Stockholm—some 9,000 or so—to dedicate its Saturday service to the laureates who are Jewish. And on this Saturday, which happens to be the first day of Chanukah, two of the laureates and their guests assemble at the two-hundred-year-old synagogue, a few blocks from the Grand Hotel. Outside, as before many such buildings in Europe, a precautionary police van is parked. Inside, the first candle has been lit.

This is a traditional congregation, with the women seated upstairs. As the service unrolls, a small boy who has just celebrated his thirteenth birthday reads the Torah. The rabbi advises him that one model of the good life is that exemplified by Aaron Klug, who has preceded the boy in reading the book. At the end of the ritual, the scrolls are returned to the Ark, and as the curtains are unfurled, it comes as a kind of revelation that it is the Book, the Word, that are the subject of worship in this place. As learning and scholarship are honored in the secular rituals of Friday, so the Word is honored on Saturday in this northernmost outpost of the Diaspora. The rabbi, an American from Philadelphia, gives a stunning sermon for the occasion. It is in English and is laced with apt quotes from Socrates and Gershon Cohen. He declares that the feast of light, the enterprises of arts and science as honored by the Nobel Prizes, and indeed the Sabbath itself are manifestations of God's gift of the Word: the gift of human reason. For the auditors—Swedish, American, English—some of whom have not been in touch with this religion since they themselves were thirteen, this celebration of reason seems a fitting end to the winter ceremonies.

14

THEY ALL LAUGHED AT
CHRISTOPHER COLUMBUS

An oily swell rolled in from the southeast, veiled cirrhus clouds tore through the upper air, light gusty winds played over the surface of the water, low pressure twinges were felt in his arthritic joints. . . . So, heaving-to off the Ozama River mouth, the Admiral sent ashore his senior captain, Pedro de Terreros, with a note to Governor Ovando, predicting a hurricane within two days . . . and begging the governor to keep all his ships in port and double their mooring lines. Ovando had the folly not only to disregard the request and warning, but to read the Admiral's note aloud with sarcastic comments to his heelers, who roared and rocked with laughter. . . . And the great fleet proceeded to sea that very day, as the governor had planned.

SAMUEL ELIOT MORISON, *Christopher Columbus, Mariner*

Morison is describing an episode of the fourth, and last, voyage of Columbus to the Indies in 1502. The admiral's fortunes were not at their peak; his costly trips had yielded few spices, insufficient gold, and Indian slaves who tended to die in transit. As an explorer he had found islands aplenty but not the treasures of Cathay; as the viceroy of Hispaniola (Santo Domingo), he had coped with climate but not with insurrection. His colonial politics were so inept that after his third voyage he was sent back to Castille in chains by a royal commissioner. The hero of 1492 became the suppliant of 1500; Hotspur grew old and ended as Lear.

"The lands in this part of the world, which are now under your Highnesses' sway," says Columbus, according to Morison, are richer and more extensive than those of any other Christian power, and yet, after that I had, by the Divine Will, placed them under your high and royal sovereignty, and was on the verge of bringing your Majesties into receipt of a very great and unexpected revenue . . . and with a heart full of joy, to your royal presence, victoriously to announce the news of the gold that I had discovered, I was arrested and thrown with my two brothers, loaded with irons, into a ship, stripped, and very ill treated without being allowed any appeal to justice. . . . I was twenty-eight years old when I came into your Highnesses' service, and now I have not a hair upon me that is not gray; my body is infirm, and all that was left me, as well as to my brothers, has been taken away and sold, even to the frock that I wore.

Reprieved by the monarchs, Columbus was packed off to the Indies one more time but forbidden to stop at the colony he had founded. The Admiral of the Ocean Sea was only fifty-one, but his chroniclers describe a far older man even by standards of the time. Harried by painful, deformed joints and inflamed eyes, Columbus nevertheless crossed the Atlantic with a flotilla of four vessels in twenty-one days; infirmity had not dulled his navigational powers. Despite orders, the admiral contrived a detour to Santo Domingo in the course of his search for a passage between Cuba and South America. Columbus was persuaded that Cuba was a peninsula of China and was confident that one more voyage would disclose a sea route to the Indian Ocean just north of Venezuela, which he had discovered in 1498.

Columbus dropped anchor off Santo Domingo, but was refused permission either to land or to put into harbor by Governor Don Nicolás de Ovando, who had been sent from Spain as the admiral's replacement. As his ships lay offshore, Columbus sensed an approaching storm by a change in weather, that low pressure and rising humidity which precede a tropical hurricane. He realized that there was ample time to issue a warning, and generously dispatched a messenger to the port. The governor, however, refused to heed the warnings of Columbus and his arthritis, and consequently Ovando's fleet was seabound for Spain on open waters the next day when the hurricane struck.

The rich armada was devastated by the storm, which sank twenty-nine of thirty ships, drowned five hundred sailors, and sent several million dollars' worth of gold bullion to the ocean floor. Meanwhile, Columbus looked after his four vessels with his usual skill. Placing his ships in the lee of a nearby river valley, the admiral was able to ride out the furious hurricane by pluck and seamanship.

"The tempest was terrible through the night," wrote Columbus, "all the ships were separated, and each one driven to the last extremity without hope of anything but death; each of them also looked upon the loss of the rest as a matter of certainty. What man was ever born, not even excepting Job, who would not have been ready to die of despair . . . refused permission either to land or to put into harbor on the shores which by God's mercy I had gained for Spain sweating blood?"

The storm not only demolished Ovando's grand fleet but drowned the royal commissioner who two years earlier had sent the admiral home in chains; it also leveled the town of Santo Domingo. No one laughed at Christopher Columbus after that storm.

In fact, as the biographers of Columbus teach us, Ira Gershwin had it somewhat wrong when he wrote:

> They all laughed at Christopher Columbus
> When he said the world was round . . .

All the chronicles agree that most European navigators were sufficiently persuaded of this fact, certainly in theory, before 1492. But though it might be amusing to see what rhymes the Gershwins could have made of such unpromising nouns as "hurricane" or "arthritis," we can be grateful that they did not succumb to historical details. Their task would not have been easy, considering that when "they all laughed at Christopher Columbus" they laughed at a sailor with Reiter's syndrome. (Try squeezing that one into a popular ballad.) But Reiter's syndrome is probably the diagnosis of the disease that plagued Columbus for the last ten years of his life, the malady that reduced the confident navigator of 1492 to the crippled Job of his last voyage. Indeed, it is likely that Columbus became ill because his genes lost a game of molecular roulette to a bacillus common in the tropics: *Shigella flexneri*.

Reiter's syndrome is diagnosed by the triad of arthritis, uveitis, and urethritis: inflammation of the joints, eyes, and terminal urinary tract. It owes its eponym to Professor Hans Reiter, who described the tribulations of a Prussian officer in the summer of 1916. The lieutenant developed painful, febrile diarrhea that lasted for two days and was followed by a week-long latent period, after which he developed painful arthritis of many joints. This episode remitted after several days, only to return; on this occasion the painful joints were accompanied by uveitis and urethritis. Reiter reported the case and speculated that it was due to a specific "spirochetal" infection of the joints. Almost simultaneously, the other side in the Great War provided evidence that the syndrome resulted from lapses of military hygiene. Doctors Fiessinger and Leroy reported in a Parisian medical journal that inflammation of joints, eyes, and the urinary tract followed an outbreak of dysentery among French soldiers on the Somme.

But, perhaps appropriately, it is from American sources that we can learn the most about the disease that crippled Christopher Columbus. If the finest history of the voyages of Columbus (*Admiral of the Ocean Sea: A Life of Christopher Columbus*) was written by our preeminent naval historian, Samuel Eliot Morison, we might also claim that the best account of the natural history of Reiter's syndrome has been given by Commander H. Rolf Noer of the U.S. Navy. Writing in the *Journal of the American Medical Association* in 1966, Noer describes a remarkable experiment of nature. In June 1962 a cruiser of our Mediterranean fleet visited a port "in a locale known to be endemic for *Shigellosis*"; naval discretion withholds the locale. A picnic had been prepared on shore in celebration of the ship's anniversary, and since this was to be a crew's party, officers did not attend. Despite rigorous sanitary precautions, the food became contaminated. It seems that two cooks, who handled all the food, had contracted mild dysentery on shore. They failed to report their illness, fearing they "might lose their promised liberty at the conclusion of the picnic—and it was the ship's last night in port. Consequently, their work of preparing food for service upon the cafeteria-style serving line was interspersed by repeated hurried trips to the toilet. They dared not take time for hand-washing lest their absences be noted."

Disaster struck within eighteen hours. As the ship was already out to sea and out of range of modern medical services, Noer and his staff were forced to cope with over five hundred cases of bacillary dysentery by their own efforts; over half the afflicted were admitted to sick bay. By the time the ship arrived at a position within reach of aid, it was clear that the worst was already over. All told, 602 cases of dysentery were identified and nine of these developed Reiter's syndrome. None of an equally large group escaping dysentery developed the syndrome. As would be predicted, officers were in the main unaffected, but one of these had picked up dysentery in a hotel and, along with the nine crew members, developed Reiter's syndrome. Noer describes the typical case as that of a sailor, who, ten or so days after recovering from bacillary dysentery, develops inflamed eyes and arthritis with an accompanying urethritis that is milder and usually not troublesome. The arthritis is "chiefly of the lower extremities, dwelling upon one particular joint."

Ten years later, two Stanford rheumatologists, A. Calin and J. F. Fries, published the follow-up of Noer's cases. By 1976, it had become clear that Reiter's syndrome was the result of a unique genetic predisposition to an environmental insult. Modern immunogenetics has detailed how the signals of our identity are displayed on the surfaces of cells by molecular markers encoded on our sixth chromosome; genetic self is distinguished from genetic nonself by means of these molecules. Follow-up studies of Finnish victims of a shigella epidemic had shown that over 80 percent of patients with Reiter's syndrome displayed the genetic marker HLA B-27. But it was not clear how this marker related to the severity of the disease over the long haul. Calin and Fries were able to trace five of the original victims on the cruiser, including the lone officer. Four of the five showed evidence of "persistent and aggressive disease," including crippling arthritis and even blindness; each was B-27 positive. The one sailor who was found to be symptom-free for two years was B-27 negative.

Calin and Fries describe what happened to these men in the decade since the onset of their disease. The "cases" are numbered as in the original report of Noer, with ages as of 1976:

> Case 4 (present age, 31) has a persistent arthritis of shoulders, elbows, wrist, hips, knees and ankles. He has severe spondylitis with involvement of sacroiliac joints and spine, including neck. He has had recurrent episodes of balanitis (inflammation of the penis) and urethritis and recently has developed blurred vision of the right eye, after recurrent acute episodes of uveitis. His clinical state has rendered him unemployable.
>
> Case 6 (present age, 31) has active arthritis of knees and ankles, persistent urethritis, and bilateral conjunctivitis and uveitis.
>
> Case 7 (present age, 36) is blind in his left eye. This has resulted from recurrent uveitis, submacular and subretinal hemorrhages, and recent retinal scarring. He is now developing acute symptoms in his right eye, associated with blurred vision. He has widespread joint involvement with limited range of movement in his left wrist and right elbow. Urethritis and balanitis have been recurrent during the last 13 years.
>
> Case 10 (present age, 34) has had recurrent joint problems, with arthritis of the left wrist and both ankles, and a history of low back pain and stiffness. Clinical examination shows a limited range of movement of his spine. . . . He has unilateral uveitis and occasional episodes of blurred vision. Balanitis and urethritis persist.

Now let us turn to the malady of Columbus. I have constructed a case history from secondary sources, so it is entirely possible that this reconstruction can be overthrown by more direct evidence. Moreover, it is impossible to determine whether the admiral suffered from disorders of gut or sex, owing to reticence in these matters on the part of his chroniclers. The case history is written from the vantage of 1506:

Case C (present age, 55). This retired admiral became ill on his second voyage to the Indies. His base of operations in Hispaniola was burdened by endemic intestinal disease to which half the colonists succumbed, forcing the admiral's doctor to test all provisions on dogs before permitting Spaniards to eat. At sea between Santo Domingo and Puerto Rico on September 25, 1494, the admiral became febrile, confused, and developed severe arthritis of his lower limbs, which was diagnosed by the local doctor as gout. He was carried ashore in the arms of his lieutenants. Bedridden for weeks, he was soon well enough to march into the interior of the island. Asymptomatic during the return from his second voyage, he fell sick again on the outward leg of his third voyage. His eyes were so inflamed that he described them as bloody. On August 4, 1498, while off the coast of South America, the admiral writes that although "blind for some time on the voyage which I had made to discover the continent, my eyes were not then so sore and did not bleed or give me such great discomfort as now." By August 13, inflammation of his eyes and recurrent arthritis were so severe that he was unable to disembark on the mainland and sent the faithful de Terreros to claim possession of the continent for Spain. Arthritis troubled the admiral throughout his colonial administration in Hispaniola; he was in a remission when remanded back to Spain. By the fourth voyage, in 1502, his arthritis had become so disabling that his men were forced to build a "kind of doghouse" on the afterdeck to save the admiral trips to his quarters. He was partially blind at times. During that final voyage, in the course of which he reached the site of the present Panama Canal, he was frequently bedridden. Years later, his brother Bartholomew recalled standing in for the crippled admiral, when "with trumpets and with flying banners [I] took possession in the name of the king and queen, our lords, because the said Admiral Christopher Columbus was indisposed at the time and unable to do so." Afflictions of the eyes and joints, as well as general debility, caused errors of navigation which led to marooning of his flotilla on Jamaica for a year. After Columbus returned to Spain for the last time in 1504, he was no longer able to travel over the rough roads of the day, not even for Isabella's funeral. Nevertheless, he was determined to settle his affairs with the king, for enemies at court had conspired to deny Columbus titles and favors. Because of spine and back pain, he first arranged for transportation by catafalque (a flat-bedded hearse), but changed his mind. Eventually he obtained special permission to ride to court on a mule, since the jostling gait of a horse would have been too much for his spine. All observers agree he died "with crippling arthritis" in 1506.

When this case history is presented to rheumatologists, they almost uniformly put Reiter's syndrome at the top of their list of possible diagnoses. That list includes a variety of other rheumatic conditions, but none rings quite so true as Reiter's. This diagnosis is surely more helpful than the account of the admiral's disease given in the most recent biography, *Christopher Columbus: The Dream and the Obsession* by Gianni Granzotto. Describing the admiral's first, febrile episode of 1494, Granzotto explains:

> Such were the effects of the disease that would accompany him for the rest of his life, a form of gout or podagra aggravated by the widespread rheumatic affections and an overall change in his metabolism. Gregorio Maranon, a Spanish authority on rheumatism and all related afflictions, made a careful scientific study of the Columbus case. He defined Columbus as the "martial" type, because of the congestion that often turned his face the color of Mars, and the helmet of prematurely white hair that adorned his head.

Perhaps because he was unable to avail himself of the expert advice of Dr. Maranon, Morison is content to inform us that Columbus was "suffering from arthritis and inflamed eyes." Some historians I have read seem to have been led astray by references to gout (most arthritis of the feet was called gout in the fifteenth century), while others speculate that the admiral had syphilis (which, we are informed, can be caught from privies!). By and large, I rather think that Columbus had Reiter's syndrome.

We are assured by all the textbooks that Reiter's syndrome is not only acquired from such unsavory intestinal bacteria as *Shigella*, *Salmonella*, and *Yersinia*, but also by venereal spread. However, much of the evidence for sexual transmission is very indirect indeed. The disease is common among young men in groups, but it is by no means obvious by what routes it is commonly transmitted; AIDS should have taught us that we frequently ask the wrong questions in this regard. For what it's worth, my personal experience suggests that the military connection is more than casual. I saw my first patient with Reiter's syndrome in the 1950s; he was a newsdealer whose disease began while he was serving with the Abraham Lincoln Brigade in the Spanish Civil War. Sixteen years after his first episode of arthritis in the trenches before Madrid, his spine had become bent, his ankles were gone, and he had lost the sight of one eye. In the slightly more sanitary setting of the U. S. Army hospital at Fort Dix a few years later, acute Reiter's syndrome was by no means an uncommon diagnosis. My friends, who at the time attended to fast-track officers at the Pentagon, assure me that the syndrome was perhaps the most common form of arthritis in patients under forty.

More recently, I see men with Reiter's syndrome at VA hospitals around the country, where the disease seems not to discriminate among soldiers, sailors, and marines. Women and blacks are rarely affected; the B-27 marker is less frequent in non-Caucasians. One memorable patient at Bellevue had no military connection. A male disco dancer, à la John Travolta, he had recurrent arthritis of both feet which became acute every Monday morning. Following his early Sunday morning exertions on the dance floor, he took to bed for the remainder of the day. When time came for work, his arthritis flared: clearly a case of Saturday Night Fever/Monday Morning Reiter's!

So much for the disease; does it permit us to understand anything at all about Columbus? Nowadays, when historians avoid traditional narratives of men and events to recount instead the sensibilities of groups, there would seem to be no great reason for unearthing the clinical details of an individual life, no matter how instructive the detail or prominent the life. But it might be argued that in the case of Columbus, these considerations are moot; the Americas were not discovered simply to expand Iberian trade or to propagate the Christian faith, although both purposes were satisfied. Freud had it right when he told Marie Bonaparte that Columbus was an adventurer whose adventure succeeded because of character. That character permitted Columbus to deal with tempest and shipwreck, betrayal and disease. Near-mutiny did not disrupt the first voyage, chains could not hold him from the last: Reiter's syndrome did not prevent Colombus from claiming new continents for Spain. As doctors learn daily from far humbler patients, diagnosis is not destiny.

The admiral died at the age of fifty-five; we do not know whether he died of one of the rare complications of Reiter's syndrome, aortic insufficiency, or whether his death was due to unrelated causes. He was accorded no state funeral; nor had he retained the powers of viceroy and governor at the time of his death. He died with his legacy in dispute. His fortunes had not significantly improved since he complained of disfavor to his monarchs:

> Seven years passed in discussion and nine in execution, the Indies discovered, wealth and renown for Spain and great increase to God and to His Church. And I have arrived at and am in such condition that there is no one so vile but thinks he may insult me. What have I not endured! Three voyages undertaken and brought to success against all who would gainsay me; islands and a mainland to the south discovered; pearls, gold, and in spite of all, after a thousand struggles with the world and having withstood them all, neither arms nor counsels availed. . . .

The continents he discovered were mistakenly named after a Florentine who cooked the data (Amerigo Vespucci deliberately predated his accounts of voyages to South America). Three centuries of litigation at court were required before his heirs took final possession of their depleted legacy. Ferdinand and Isabella had promised their Admiral of the Ocean Sea heritable dominion of all he might discover. The admiral's twentieth-century descendant, also named Christopher Columbus, held title only to the dukedoms of Veragua and La Vega (bits of Central America and a valley in the Dominican Republic); he was killed by Basque nationalists.

Freud was persuaded that Columbus was no genius (*grosse Geist*), but I think another doctor, William Carlos Williams, came closer to the mark when he wrote in *In the American Grain*:

> As much as many another more successful, everything that is holy, brave or of whatever worth there is in a man was contained in that body. Let it have been as genius that he made his first great voyage, possessed of that streamlike human purity of purpose called by that name—it was still as a man that he would bite the bitter fruit that Nature would offer him. He was poisoned and his fellows turned against him like wild beasts.

The story of Columbus separates itself from the broad outlines of historical narrative to fit into the stricter templates of tragedy. His "human purity of purpose" was interpreted as overweening pride; his downfall evokes pity and terror not only in Dr. Williams. But the Columbus who reigns in our imagination has not been poisoned by disease or brought low by courtiers. His is the name we associate with the experience of the new, whether it be the landscape of the moon or the cartography of the sixth chromosome of man. When we open the doors of our laboratories each morning, we start on voyages of discovery, which in our wildest fancies might lead to descriptions like this: "Presently they saw naked people," writes Morison,

> and the Admiral went ashore in the armed ship's boat with the royal standard displayed. So did the captains of *Pinta* and *Niña* . . . in their boats, with the banners of the Expedition, on which were depicted a green cross with an F on one arm and a Y on the other, and over each his or her crown. And, all having rendered thanks to Our Lord kneeling on the ground, embracing it with tears of joy for the immeasurable mercy of having reached it, the Admiral arose and gave the island the name of San Salvador. Thereupon he summoned to him the two captains, Rodrigo de Escobedo, secretary of the Armada, and Rodrigo Sánchez of Segovia, and all others who came ashore, as witnesses; and in the presence of many natives of that land assembled together, took possession of that island in the name of the Catholic Sovereigns with appropriate words and ceremonies.

15

WESTWARD THE
COURSE OF EMPIRE

The Upper West Side of Manhattan, where I have lived since childhood, is guarded by four statues of reassuring virtue. Its southern boundary is marked by the figure of Christopher Columbus on a tall column in the middle of the busy traffic circle that bears his name. The column is decorated with galleon-like ornaments invisible to the admiral from his high perch. He stands erect and faces south so that his eyes will avoid forever the porn houses of Eighth Avenue to fix on the silver harbor in the distance. Some fifty blocks to the north sits Columbia's *Alma Mater*. From her throne on a piazza before the rotunda of Low Library she rules the university with the graceful scepter of learning. Her sculptor, Daniel Chester French, has hidden the owl of Minerva in the folds of her skirt: woman and bird look south over our neighborhood from Morningside Heights.

Our western boundary is guarded by an equestrienne. From a hill at Riverside Park and Ninety-third Street, Joan of Arc flashes her unsheathed sword westward at New Jersey across the Hudson. Triumphant in armor, the warrior saint leans back in the saddle to hear voices of children at skirmish in the grass. Also a mounted figure, the largest sculpture to defend our border is on Central Park West, our eastern limit. Before the American Museum of Natural History rides the hero of San Juan Hill, Teddy Roosevelt. Dressed in frontier garb and mounted on a great stallion, he is flanked by two noble savages on foot, an Indian and a black, who are prepared to follow the rugged president on his eastbound charge across the park to pacify Fifth Avenue.

Although the subjects of these monuments cannot have been chosen with an eye for the fancies of civic allegory, one might still read from them a lesson in urban affairs. The statues of three of our four guardians—

Columbus, Saint Joan, and Teddy Roosevelt—have been placed with their backs to the district, while only Minerva and her owl study the Upper West Side. Discovery, Faith, and Enterprise look forever at the far horizon; Wisdom watches the neighborhood. It is a neighborhood that has undergone a remarkable transformation in the last twenty or so years, and the lessons to be drawn from that cycle of transformation appear to me to apply not only to urban affairs, but to other endeavors as well: medicine, for one.

The first lesson is summed up by the aphorism of Lewis Mumford: "Trend is not destiny." And no one will dispute that the trendiest street of the West Side in the 1980s has been Columbus Avenue. This newly fashionable street runs north and south more or less from Columbus Circle to Columbia University. Twenty-odd years ago the trend was elsewhere. Columbus Avenue was then a disaster area, a nighttime playground for street gangs and a daytime rendezvous for child molesters, which urban planner Jane Jacobs described in her 1961 book *The Death and Life of Great American Cities* as:

> endless stores and a depressing predominance of commercial standardization. In this neighborhood there is so little street frontage on which commerce can live, that it must all be consolidated, regardless of its type.... Around about stretch the dismally long strips of monotony and darkness—the Great Blight of Dullness, with an abrupt garish gash at long intervals. This is a typical arrangement for areas of city failure.

But this area of the city was not destined to fail; Jacobs was simply describing a trend that—ironically—her book helped to change, creating the little renaissance of our neighborhood. *The Death and Life of Great American Cities* was a call to arms in the fight against urban renewal as expressed in mass demolition. Jacobs, and her contemporaries in urban sociology, pointed out that the utopian sketches of Ebenezer Howard's *Garden Cities of Tomorrow* or Le Corbusier's *La Ville radieuse* earlier in the century led to the bulldozing of vital parts of such substantial cities as Boston, Philadelphia, Chicago, and New York. They argued that city planners of Howard's persuasion were motivated by antiurban prejudices. A product of the English pastoral tradition, Ebenezer Howard was certain that good stiff doses of country air and green grass would serve as antidotes to the diseases of urban crowding, pestilence, and crime. His solution was to disperse the city dwellers in the fresh countryside.

Howard's views were taken up in the United States by a group called the Decentrists, among them Lewis Mumford, Clarence Stein, and Catherine Bauer. As a student, I remember reading Lewis Mumford's *The Culture of Cities* on the eighth floor of a snug Manhattan apartment house; he persuaded me at the time that life in cities-as-they-are would soon become intolerable. Along with Siegfried Gidieon, an architectural historian prominent in the 1940s, the Decentrists argued that cities of the future would survive only if planners were permitted to separate commerce from living quarters, automobiles from pedestrians, work from leisure: one area for one use. Their aim was either to depopulate the center of cities or to bring the parkways into downtown. Indeed, if one believed—with Catherine Bauer—that the cities presented "a foreground of noise, beggars, souvenirs and shrill, competitive advertising," one might also agree with Gidieon that "there is no longer any place for the street with its traffic lane running between rows of houses; it cannot possibly be permitted to exist. . . . Then the parkway will go through the city as it does today through the landscape, as flexible and informal as the plan of the American home itself."

The parkway solution, in which wide swaths of roads separated gardenlike suburbs of super-blocks, was expressed in its popular form by the diorama that General Motors mounted at the World's Fair of 1938–39. This toy utopia won a generation of converts to the theology of freeways. E. L. Doctorow recalls the "World of the Future" in his novel-cum-memoir, *World's Fair*:

> This miniature world demonstrated how everything was planned, people lived in these streamlined curvilinear buildings, each of them accommodating the population of a small town and holding all the things, school, food stores, laundries, movies and so on, that they might need, and they wouldn't even have to go outside, just as if 174th Street and all the neighborhood around were packed into one giant building.

In that vision, the simple street had vanished, differentiating itself into superhighways for cars and walkways for pedestrians. Apartment houses and old brownstones disappeared into multistory ziggurats, and the social life of the street was left to take care of itself on grassy meadows between high-rises.

In the course of the immense building boom after the Second World War, many of the goals of Le Corbusier and Gidieon were realized. Older buildings were demolished, freeways broke their way into the center of lively neighborhoods, and instant cities of the future were precipitated in desert or jungle. The result of this expansive construction, the outcome of the old bias against the city-as-it-is in favor of the city-of-tomorrow, can be seen all over the globe: from Los Angeles to Moscow, Houston to Belgrade, Brasília to Chandigarh. In Manhattan, high-rises of red brick were built as a kind of giant picket fence to separate Harlem from midtown, the bustling Lower East Side near the Brooklyn Bridge was leveled to build a web of twenty-story fortresses, and the Upper West Side was threatened by "urban renewal" at both ends.

But this trend was not to be our destiny, thanks in part to the clear diagnosis spelled out by Jane Jacobs. She not only identified the problem but also suggested its solution. If the real sickness of the city was a Great Blight of Dullness, then the cure was Diversity: one area for many uses. A diverse neighborhood, a district "big enough to fight city hall," could save its old housing stock, watch the streets, and regain control of its own turf. Only a good-sized district, diverse in population and commercial activity, could protect itself against hoodlums from within and speculators from without.

Jacobs detailed how many such neighborhoods came back from the slums in this way: Chicago's Back of the Yards area, Philadelphia's Rittenhouse Square, Boston's North End, New York's West Village. Her clinical insight into the process of "deslumming" led her to a rather unique diagnosis. (The medical analogy may not be inappropriate, since Ms. Jacobs laces her argument with examples from biomedical discourse.) Reasoning that diversity was the heart of the urban matter, she examined the natural history of deslumming to identify a "generator of diversity." Her search yielded the observation, novel for the time, that neighborhoods are reborn in the presence of a vigorous and independent commerce.

The quest for a "generator of diversity," with its evocative acronym, suggests another analogy to medical science. Perhaps by coincidence, immunologists in 1960 or thereabouts also embarked on a search for a "generator of diversity" to solve the riddle of antibody formation. If one gene made one protein, how could our antibody-forming cells possibly have sufficient genetic information to make antibodies to the uncountable antigens we might meet? Some of these antigens do not even exist in the natural world. We are, nevertheless, endowed with the capacity to make antibodies to molecules still fresh from the drawing boards of industry. This preknowledge, so to speak, accounts for some allergic reactions to drugs; but how can our genes have been programmed to recognize those new chemical structures? The successful solution of this riddle culminated in the discovery of gene rearrangements in the course of B-cell development. It was learned that cells trade bits and pieces of old genes to make new ones. The cells that make well-adapted new genes are those which are selected for growth and development. Urban planners of Jacobs' stripe will understand biologists who describe antibody formation in language such as this: "Thus overall, we may view the capabilities of the immune response as being derived predominantly from selective forces, where extent of diversity and adaptability are the most important features."*

When Jacobs identified diversity as the goal toward which commerce is the goad, she anticipated the modified Darwinism of modern biology; Manser's "diversity and adaptability" are achieved by the vigorous commerce of genetic units. Jacobs' entrepreneurial bent was surprising at the time since the literature of city planning had for so long been the bailiwick of dreamy utopians. The words are the words of Jacobs, but the hand is the invisible hand of Adam Smith:

> Most of the uses of diversity . . . depend directly or indirectly upon the presence of plentiful, convenient, diverse city commerce . . . whenever we find a city district with an exuberant variety and plenty in its commerce, we are apt to find that it contains a good many other kinds of diversity also, including variety of cultural opportunities, variety of scenes, and a great variety in its population and other users.

In other words, when you've got them by the shops, their hearts and minds will follow. The rise of Columbus Avenue illustrates this principle. In the early 1960s, only a few stores in our immediate neighborhood had survived the postwar flight of the middle classes to suburbs on the parkways. The clientele of those small service stores were mainly aging widows of Central Park West or a few young professionals, a class in those days not as affluent as now. Morris, the candy store owner, sold newspapers, art supplies, and *Partisan Review* in a shop next to an abandoned meat market. Demetrios, the shoemaker, fixed our dog's leash for a quarter and sold laces for the kids' ice skates. Alan, the druggist, sold trusses and notarized without pay the many petitions we and our neighbors sent to city hall.

*Manser, T., et al. *Immunology Today* 6: 1–7, 1985.

Our block was no radiant city in those days; the apartment building in which we lived was next to a large single-room-occupancy hotel inhabited by a villainous crew. Drug addicts and dealers, pickpockets and muggers, hookers and pimps kept their fellow hotel guests in constant terror. At one of our corners was a bar frequented by men who would lurch into the street either drunken or stabbed; at another corner was a greasy spoon where derelicts sat for hours over tepid soup. There was in fact no clean or safe place to eat in the neighborhood, and after the shops shut in the evening, the streets were ready for the melodramas of city crime. I will never forget looking out one cold night in December from my window on Seventy-seventh Street to see two coatless men in white shirts running out of the bar into the glare of streetlights; one had a gun in his hand, which he proceeded to empty into the chest of the other. The victim's shirt turned as red as the poinsettias in our lobby. Lower-level crime was more common: the half-life of our children's bikes was five months, and our Chevy was always fair game for petty larceny. The only store that prospered in those days was the locksmith's.

We stuck it out for all the conventional reasons that city dwellers give themselves for tolerating squalor: convenience of the area to work, the hassle of commuting from greener pastures, culture, continuity, and finally, diversity. This we found aplenty: diversity of class, color, dress, generation, and belief. And with time the squalor dissolved, probably for the reasons advanced by Jane Jacobs: "Being human is itself difficult, and therefore all kinds of settlements (except dream cities) have problems. Big cities have difficulties in abundance, because they have people in abundance. But vital cities are not helpless to combat even the most difficult of problems."

The chief impetus for change on the West Side was the construction of Lincoln Center, a monument to the Camelot era. And the first harbinger of change was the opening of a few good pubs and restaurants in the neighborhood of Lincoln Center. The new eateries brought people into the neighborhood in the evening. Hamburgers and eggs Benedict gave way to tacos, linguini, and sushi as cheap ethnic bistros opened up to catch the overflow of evening diners. The streets became full as not only the culture vultures of Lincoln Center but locals themselves nipped off around the corner for a bite. New York City helped by easing the restrictions on sidewalk cafés, which soon ranged from Anita's Chili Parlor to the glassed-in splendors of the Museum Café. This establishment, across the Avenue from the Museum of Natural History, remained for a short time the northernmost outpost of Quicheland; on the day that the Museum Café replaced our corner crime bar, it was clear to one and all that gentrification had arrived. The greasy spoon at our other corner was successively transformed into (a) a modest natural-food parlor, (b) a trendy singles café featured on the cover of *New York* magazine, (c) the temporary home of Paul Prud-homme's Cajun cuisine, and (d) a button-down dining establishment where one person can eat Sunday brunch for fifty dollars.

The crowds in the eateries and at sidewalk tables made for a safe Avenue, and in turn a thousand retail shops bloomed. For a short and glorious time in the early 1970s there must have opened a dozen antique shops where vintage clothing, movie posters, unfinished pine chests, and Bakelite jewelry reentered the vernacular of modern life. Soon the Avenue was given over to aspects of the singles scene, as migrants from suburbs or Akron found the West Side suitably safe for a weekend stroll in pursuit of wicker, quiche, or company.

One more element entered this mix. As more and more women worked at jobs and for wages the equal of men's, it followed that take-out food replaced much of the home cooking they had earlier provided. In any event, it took less than a decade for the neighborhood to be turned into a little souk for gourmets, from which breadwinners of either gender could return with upscale snacks for the evening meal—truffled pâtés and stuffed em-panadas, honeyed quail and saffroned rice, barbecued ribs and golden couscous. Nor did the customers have a long walk home, because the West Side had become the hot growth center of Manhattan real estate.

Between 1965 and 1985, most of the brownstones on the side streets and all of the Avenue's tenements were renovated, the welfare hotels closed, and the older apartment houses spruced up as tenants became owners rather than renters. New dwellers meant new needs, and since many of the new inhabitants were young and affluent, specialized shops offering nonessentials of all kinds were favored. And beneath the bushes of smart boutiques grew the weeds of card shops and Tofutti parlors. If the area supported only one or two bookstores but ten outlets for video rentals, if it boasted of only one art gallery but fifty shoe shops, it had nevertheless achieved in two decades all that Jacobs could have wished in the form of diversity. And the streets had become safe.

The second lesson is not from Mumford, but from Jacobs. Discussing the decline of Eighth Street in Greenwich Village, she formulated the principle that "diversity can self-destruct." In a passage that seems as

pertinent to the Columbus Avenue of today as to the dilemma of the Village
in 1960 she recounts that:

> Among all the enterprises of Eighth Street, it happened that restaurants
> became the largest money-earners per square foot of space. Naturally
> it followed that Eighth Street went more and more to restaurants.
> Meantime, at [one] corner, a diversity of clubs, galleries and some small
> offices were crowded out by blank monolithic, very high-rent apart-
> ments. [One] watched new ideas starting up in other streets, and fewer
> new ideas coming to Eighth Street . . . if the process ran its full and
> logical course, Eighth Street would eventually be left beached, in the
> wake of popularity that had moved away.

Absent intervention, Jacobs' principle of the "self-destruction of diversity" operates like clockwork. If no one intervenes, rents rise to meet the
profit generated by the most profitable enterprise. The clock is ticking away
in our less and less diverse neighborhood for the small shopkeepers who
cannot compete with restaurants and chain stores.

And sure enough, Morris is gone, his site preempted by a boutique with
a branch near Bloomingdale's. Demetrios has been replaced by a kinky shoe
store in the windows of which manikins are arranged to illustrate that
podology recapitulates misogyny. Alan has moved off the Avenue and in
his place a Yokohama jewelry store has plunked its New York affiliate. The
large retail chains of Europe moved in in rapid pursuit of trendy positions
on the Avenue as gentrification progressed northward past the Museum.
When Laura Ashley or Crabtree & Evelyn established their genteel colonies,
we did not ask for whom the bill tolled. But when the first of our Benetton
sweater shops appeared, the prognosis seemed as ominous as the first
dusky spot of Kaposi's sarcoma. Predictably, the sweaters have metasta-
sized. The lineup on Columbus Avenue is beginning to look like the tax-free
shops at Charles de Gaulle Airport.

Protected from hoodlums by gentrification, we seem to be rolling over
for the developers. Huge high-rises have sprung up to dwarf our precincts;
at sunset the shadow of a giant postmodern tenement dominates the
Hayden Planetarium. The Trump Organization, the likes of which have
ruined more urban acreage than the Luftwaffe, promises, according to an
advertisement appearing in our neighborhood paper, *The Westsider*, to
develop at our western flank by the river "one hundred acres of new homes,
shops, restaurants, parks and promenades . . . it will be the most extensive

reclamation and development project ever attempted anywhere in the world."

The ad does not mention that the development will be called Television City, nor does it mention that it may include the tallest building in the world. This self-serving bit of public relations reminds us of Jacobs' prediction for Eighth Street, that "the worst potential threat to its diversity and its long-term success is, in short, the force let loose by outstanding success." Good-bye, Columbus.

But the Jacobs' principle, the self-destruction of diversity, affects other areas of activity as well. In *The Westsider* I also note, with disappointment, evidence of the self-destruction of *medical* diversity. Two specialists in periodontia on Central Park West promise the cure of gum disease. Illustrated by a frontal shot of a student-level Nikon microscope, their ad promotes "Alternative Nonsurgical Treatment for Gum Disease" and urges prospective clients to come in for a "microscopic examination of plaque from under your gums." To my knowledge, since the etiology of periodontal disease is as obscure as that of rheumatoid arthritis, this examination offers the laity an explanation not yet available in biomedical literature—we know that bacteria are involved, but not what factors in the host permit the disease. Adjacent to this claim for new therapeutics is an equally under-stated advertisement inserted by a board-certified dermatologist with offices on Fifty-seventh Street that are "modern and conveniently located." Available on weekdays, and with "weekend hours," he is prepared to accept most insurance programs and will treat not only "diseases of skin" but acne, nails [*sic*], hair disorders, and perform "cosmetic surgery."

Perhaps because of the lower rates of remuneration in his specialty, a board-certified psychiatrist (no address, just a telephone number) is forced to advertise in the classified columns immediately above the New York Tattoo Club. The shrink offers "COGNITIVE THERAPY, Psychotherapy, direct and to the point. Especially effective in treatment of depression."

I wonder whether I should call the doctor about *my* depression; I'm depressed that our profession has in its ranks beneficiaries of four years of college education, four years of difficult study in basic and clinical science, and at least four more years of extensive specialty training, who because there are too many of them in their overspecialized field are forced to troll for patients in classified ads among masseurs. When my father and his prewar generation trod the West Side with their small black bags on house calls, they may have had a backlist of uncollected bills, they may have charged two dollars for a home visit and no extra to look at the smear from a diphtheritic throat, but they were never reduced to peddling their trade in the want ads between exterminators and pet groomers.

Another symptom of the self-destruction of diversity is displayed at the entrance to the former welfare hotel next door. The junkies and whores are long gone, displaced by energetic yuppies who spend what they call "more than a K a month" for studio apartments. The ground floor of this dormitory is devoted to a health maintenance organization, the appointments of which resemble one of Colonel Sanders' many parlors. The shingles at the door proclaim the names of the doctors and their (sub)specialties. I note three bona fide internists, two obstetricians, a surgeon, two pediatric endocrinologists, and one pediatric oncologist; there are six radiologists. It is good to be reassured that the two West Side babies unfortunate enough to develop adrenal tumors in the next five years will have a proper roentgenologic evaluation.

If these then are the symptoms, what is the diagnosis, and how does it relate to the death and rise of Columbus Avenue? The inequities of medical care ranked high among the social injustices that the radicals of the 1960s brought to our attention. The maldistribution of physicians, the impediments of access to clinics or hospitals, the rising costs and tangled referral patterns constituted a slumlike cityscape of medicine. Since the problem was perceived to arise from an insufficiency of doctors as well as their absence from ghetto and farm, a solution was found that Ebenezer Howard would have approved. We simply proceeded to double the number of our graduates, based on utopian projections of the need for physicians. To this end, new schools were precipitated in town and country, while the older schools expanded. The free-market economists believed that overcrowding the profession would lower prices; the progressives believed that overcrowding, like parkways, would bring doctors to the blighted inner cities or deposit them in the boonies.

But although medical decentrists created the large medicine-of-tomorrow in the form of Medicare, Medicaid, HMOs, and so on, our young graduates opted for diversity. Instead of *leaving* the teaching hospitals and cities in which they worked, they elected to remain for further training. They knew all too well that when the planners had their day the race would go to the best certified. And so a thousand subspecialties bloomed. We now have a generation of diversely trained pediatric ophthalmologists, adolescent urologists, geriatric cardiologists, and so forth. Perhaps never before in the history of our craft have we taught so many so much about so little.

At the technical level, no doubt, the diversity of subspecialists guarantees that nowhere else in the world is the craft better practiced. One remembers a foreword by Robert Graves, who was proud to write poems for poets: The process, he said, resembles the practice of Scilly Islanders, who in the absence of commerce take in each other's washing for a livelihood. Nowhere in the Western Hemisphere, claimed Graves, is washing better done. Be that as it may, the diversity of our specialists has had other effects. Training our fine young experts costs money and so do the machines that they require. Add those costs to the more awesome increases in the costs of hospital maintenance and administration, and one understands why in medicine as on Columbus Avenue "the worst potential threat to its diversity and its long-term success is, in short, the force let loose by its outstanding success." Pessimists in my neighborhood complain that they have survived the muggers but will not survive Donald Trump; pessimists in my hospital tell me they have survived the on-every-other-night internship, but not treatment according to "disease-related groups," or DRGs.

The third of our West Side lessons comes from Yogi Berra: "It ain't over till it's over!" and is illustrated by the best work of art on our side of town. On the walls of the New York Historical Society hang the five canvases painted by Thomas Cole in 1835–36 and titled "The Course of Empire."

The first painting shows the dawn of Empire in *The Savage State*. With the agitated strokes of Romantic painting, Cole brushes a pink sunrise over the gray mist of night. The morning light bathes a coastal inlet dominated by a steep promontory. From the forests in the foreground, a fur-clad savage has put an arrow through a deer. A hunting party emerges from a thicket in the midground; in the distance we see a clearing on which a few tepees surround a tribal fire. In this garden of the New World, only victory in the hunt distinguishes man from beast.

The second painting depicts *The Arcadian State*. Centuries have passed and the forests have been cleared. In the foreground, we see a philosopher seated on a rock, before which he traces the intersects of a perfect circle. A youth stoops in the roadway to sketch a crude figure in the sand. Where once stood the tents of savages, we now see graceful dancers on a greensward. In the midground of these Grecian scenes, we find the smoke of sacrificial fire rising from a circular temple that bears an unsettling resemblance to Stonehenge. The pastoral vision is embellished by images of sail

Thomas Cole, *The Course of Empire: The Savage State.* New-York Historical Society, New York. © Collection of the New-York Historical Society.

Thomas Cole, *The Course of Empire: The Arcadian State.* New-York Historical Society, New York. © Collection of the New-York Historical Society.

Thomas Cole, *The Course of Empire: The Consummation of Empire*. New-York Historical Society, New York. © Collection of the New-York Historical Society.

Thomas Cole, *The Course of Empire: Destruction.* New-York Historical Society, New York. © Collection of the New-York Historical Society.

Thomas Cole, *The Course of Empire: Desolation*. New-York Historical Society, New York. © Collection of the New-York Historical Society.

and harbor, plowman and field, shepherd and flock. In the bright light of midmorning, we see an even greater mountain rising in the far distance. The brushwork is finer, and the scene is bathed in blues and greens. No machine has yet disturbed our garden.

The third painting is the largest of the five. This centerpiece is entitled *The Consummation of Empire* or *The Luxurious*. Cole's palette pays homage to Turner: The picture shines in white and gold and pale orange. The inlet has now become a splendid classical harbor filled with opulent ships. Great arcades and complex temples, palaces, and colonnades cover every foot of the shoreline. The architect seems to have anticipated Albert Speer—or Michael Graves. No tree or bush survives; in the distance even the steep promontory is practically engulfed by marble terraces. The inlet is spanned by a viaduct over which a triumphant noonday procession moves to a festive arch. The great columns are bedecked with flowers, and every inch of space is crowded with an affluent populace in the grip of an obsessive leisure. There is no evidence that anyone actually *works*.

The fourth painting shows the consequences of forsaking the Arcadian state, and is called *Destruction*. A savage enemy has attacked the city and is shown in the process of sack and rape. The palette is given to grays, reds, and browns: the sky is as dark as in the savage state. The scenes of slaughter and battle are dominated by the mutilated statue of a powerful warrior in the right foreground. The classical colonnades on each side of the harbor seem to function as a marble trap which is closing on the carnage. There is no escape by land, since the people have shut themselves off from the world of nature. There is no escape by sea, since the bridges are down and the ships are sinking. Cole has shifted his point of view so that the vanishing point of the perspective leads us to the open sea, where no rescue waits.

The most conventional image in the series is the last, *Desolation*, which provides the picturesque ruin required by Romantic aesthetics. As described in the 1848 catalogue:

> The moon ascends the twilight sky near where the sun rose in the first picture. The last rays of the departed sun illumine a lovely column on the once proud city, on whose capital the heron has built her nest. The shades of evening steal over shattered and ivy-grown ruins. The steep promontory . . . still rears against the sky unmoved, unchanged.

Cole was persuaded that the function of painting was to instruct the viewer in aspects of morals. He was also persuaded that the New World could avoid the imperial fate of the Old World only by remaining at the Arcadian state. On this continent we could finally escape from the agelong cycle of rise and fall which had left romantic ruins all over Europe. Providence had chosen us for this task because of our encounter at the stage of discovery, the savage state, with the "sublime force of the wilderness." Having once seen Niagara, who would build a Babylon on the Hudson?

Cole's vision has persisted as the basis of most modern antiurban attitudes in city planning—or in medicine: Our crowded neighborhoods or ever-more-complex professions seem always on the verge of "Destruction." This Romantic notion seems to imply that cities require periodic exorcism so that their inhabitants will disperse to the Arcadian state.

But this cyclical model may be totally wrong when applied to real cities, to real neighborhoods, or to professions that move unpredictably to guarantee that "trend is not destiny." Remember when OPEC was going to eliminate the gasoline engine? Remember when you couldn't sell office space in New York or buy it in Houston? Remember when young docs were going into hyperbaric medicine?

When one considers the changes in our neighborhood—or profession—I'm not sure that it is ever possible to fit real life into the Romantic tableaux of Cole's imagination. "It ain't over till it's over" suggests that we don't *have* to let Trump build colonnades that keep us from our river. We don't *have* to spend the rest of our lives with the kind of fragmented medicine promised by subspecialization and the DRGs.

If there has been one major counterforce to the antiurban strain in our heritage it has been our experimental, technical, industrial tradition: We have put machines in our garden. Those urban machines require humans for their nurture, a requirement responsible in part for the mass immigration of the last century and a half. The urban, polyglot tradition of our democracy guarantees that local trends do not become national destiny because each group that is washed ashore has a different history, expectation, and set of skills. This tradition guarantees the kind of diversity that does not need to destroy itself. The Italians of Boston's North End, the Poles of Chicago's Back of the Yards, the Jews and Hispanics of the West Village reversed their local trends. Where once the Upper West Side took in the Irish and the Jews, it now has accommodated an influx of immigrants from the rural South, the Caribbean, the Chinas, Pakistan, India, Korea, and Russia. Our streets are alive with the sound of Babel. Look around our diverse neighborhood—or the medical schools and hospital wards—and you will see new Americans at the stage of discovery, not of effete opulence.

Immigrants arriving: That is the real generator of diversity in this country, and there is no reason at all to suppose that the acronym is inappropriate. *This* legacy—not the memory of Niagara—is the real means of escape from the cycles of imperial rise and fall. It is the constant renewal of America through the gateways of its cities that can alter the worst-case predictions of antiurban Romantics. As I bite the fresh apple bought from the Korean grocer whose shop just opened around the corner, I recall Walt Whitman's "Manahatta":

> The Countless masts, the white shore-steamers, the lighters,
> the ferry boats, the black sea steamers well model'd
> The down-town streets, the jobbers' houses of business, the
> houses of business of the ship merchants and money brokers,
> the river streets,
> Immigrants arriving, fifteen thousand in a week . . .
> Trottoirs throng'd, vehicles, Broadway, the women, the shops
> and shows
> A million people—manners free and superb—open voices—
> hospitality the most courageous and friendly young men
> City of hurried and sparkling waters! City of spires and masts!
> City nested in bays! My city.

NULLIUS IN VERBA
LUPUS AT THE ROYAL SOCIETY

Nullius in Verba reads the carved motto on the coat of arms which gleams from the rostrum of the Royal Society. A contraction from the Epistles of Horace, the phrase is there—so we are told—to remind us that in this hall truth will be tested not by words, but by experiment. Above the podium hangs the varnished portrait of Charles II, who granted the first charter to the "Royal Society of London for Improving Natural Knowledge" in 1662. The Society is housed in a cream-colored Regency building nestled in the terraces of Nash country, off Pall Mall.

But it is not the architecture which ordinarily engages the attention of a scientist on his way to the podium of the lecture theater. The halls through which he must pass are hung with the portraits of former Fellows of the Royal Society, whose names suggest standards of excellence higher than those of the usual professional fraternity. Hobbes and Locke, Pepys and Newton, Faraday and Rutherford, Darwin and Huxley, Bragg and Dale: their images seem guaranteed to humble even the most experienced scholar.

On a brisk September afternoon recently, the curtains were drawn against the autumn sun. The portrait of King Charles was eclipsed by a bright screen, upon which a succession of slides showed the astonishing scope of modern biology. At this meeting, which was devoted to fundamental aspects of rheumatology, the screen was filled with images of cloned cells and deciphered genes. Speaker after speaker—from England, Sweden, Belgium, and the United States—took a turn at displaying the latest rabbit plucked from the hat of recombinant DNA or monoclonal antibody.

I found it difficult to avoid the somewhat amused gaze of Charles' face from the intermittently lit portrait. He was monarch of the Restoration

(1660–1685), and his reign followed Oliver Cromwell's revolution of the saints (to use Michael Walzer's phrase). Charles' long black hair framed a chiseled face; he reminded me of the young Guillermo Vilas at the U.S. Open. Indeed, one of his contemporaries admired his "motions that are so easy and graceful that they do very much recommend his person when he either walks, dances, plays . . . at tennis, or rides the great horse, which are his usual exercises."

He was also fluent in French, and knew Italian and Spanish; he was deeply interested in the fashionable pseudosciences of physiognomy and astrology. More to the point, his aesthetic interests included astronomy and "experimental physics" for the study of which he maintained an amateur laboratory. I fancied that he might have been the first among us to appreciate that the optics of Isaac Newton and the lenses of Robert Hooke were directly responsible for the images which flickered from the slide projector over his painted likeness. Indeed, it occurred to me that a good portion of what was happening in that room three centuries later could be traced to Charles II and his Royal Society. It might be said that we are the children of the Revolution of the Skeptics which began under the motto *Nullius in Verba*.

One presentation seemed to illustrate this point in unambiguous fashion. For several years there has raged in the medical research literature a controversy with respect to that awful malady systemic lupus erythematosus. Several laboratories have found that patients with the disease—and many of their relatives as well—lack appropriate numbers of complement receptors on the surfaces of their blood cells. Complement receptors help the body dispose of the residues of immune reactions that arise in the course of even the most ordinary infections. A deficiency of these receptors might therefore explain why extensive deposits of immune complexes clog the blood vessels and kidneys of patients with lupus. It remained unclear—hence the controversy—whether the receptor abnormality was a primary flaw in the genes or whether this was simply one more mysterious side effect of the disease.

In the course of his presentation that afternoon, Douglas Fearon, an accomplished immunologist from Harvard, presented convincing evidence that familial variations of receptor number are associated with a characteristic arrangement of the receptor genes. One of the striking images Fearon displayed on the screen was that of a series of black and white bands representing the receptor protein, displayed directly above the matching black and white bands representing the DNA coding for that protein. Both

patterns had been obtained from the same patient, sick or well; the signifi-
cance of that observation was lost to few in the audience. Abnormalities of
receptor number in lupus erythematosus are not simply the by-product of
widespread disease, but due to *genetic* differences.

The likeness of Charles II presided over the discussion that followed;
his Society provided the tools which have made the scientific study of
disease possible. We define systemic lupus erythematosus, in part, by
fluorescence microscopy of the blood cells of our patients. This sort of
analysis we owe to the work of Robert Hooke, who not only made microscopy practical but whose observations suggested that the cell was the
elementary unit of living matter. The experiments also derive from the
work of Hooke's longtime rival, Isaac Newton, without whose optical
discoveries modern spectrophotometry and fluorescence analysis would
have been impossible. The statistics required for the discussion of genetic
data were based on methods introduced by two other early members of the
Royal Society, Sir William Petty and Captain John Graunt.

But the most direct ancestor of Fearon's discovery was the Honorable
Robert Boyle—"the Father of chemistry and the Son of the earl of Cork"—
whose high rank gave the new science its bona fides. Boyle is remembered
by every high school student, if not every pulmonary physiologist, as the
contributor of V (for volume) to the law of perfect gases ($PV = n\,RT$). But
perhaps Boyle's major achievement was to distinguish the facts of chemistry from the opinions of alchemy. It is difficult to describe DNA without
reference to its substituent purines and pyrimidines, and even more difficult to refer to these bases without reference to *their* substituent elements.
It was Boyle who finally dispensed with the old notions of "elemental"
earth, air, fire, and water, announcing in *The Skeptycal Chymist* the new,
Restoration definition of an element:

> certain primitive and simple, or perfectly unmingled bodies; which, not
> being made of any other bodies, or of one another, are the ingredients
> of which all [other] bodies are immediately compounded, and into
> which they are ultimately resolved.

Teatime, that elemental English pause, interrupted these ruminations.
We stood about the sunlit salons of the Society's gracious new quarters in
Carlton House Terrace, to which it moved in 1967 from Burlington House
in Piccadilly. As we nattered, sipping strong tea or unspeakable coffee, we
were reminded by the portraits around us that early membership in the

Royal Society was by no means limited to gentlemen of science. Among the founders of the Royal Society was the architect Christopher Wren, and its rolls numbered the diarists John Evelyn and Samuel Pepys (known to their associates as botanist and naval administrator, respectively), the poet John Dryden, the philosopher John Locke, the duke of Buckingham, the earl of Sandwich, and the Moroccan ambassador! The Society grew rapidly—there were 131 Fellows in 1663, and 228 by 1669; recent critics suggest that the group was diluted by amateur aristocrats. Election was in principle unrestricted, but in 1847 (to quote a Society pamphlet) "it was decided to limit the number of Fellows elected annually and to restrict election to those distinguished for their original scientific work."

We returned to the lecture hall, where we were plunged into the immunology of the 1980s. We were back to lupus and its unfair toll of the young. The disease afflicts more women than men by a ratio of about nine to one, many of these women being of childbearing age. There is, indeed, some evidence that hormones deeply influence the onset and course of the illness. Nowadays the outlook is far less grim for women with lupus than it was during my house-staff days when cortisone and its derivatives had just started to turn the tide. Nevertheless, the disease remains the major cause of death within my clinical subspecialty, rheumatology.

Patients with lupus make antibodies to almost every normal constituent of their own cells, as if these constituents belonged to some foreign invader that required tarring with the brush of antibody. Many of the antibodies are directed against nucleic acids, DNA and RNA; many are raised against cell membranes and structural proteins within the cell. Indeed, the novel antibodies raised in the course of this confusing disease have permitted modern molecular biologists to detect previously unrecognized components of our cells and organelles. Several of the self-directed antibodies in the serum of patients with lupus can be used to help determine the diagnosis and outcome of the disease; unfortunately, they also play an important role in its progression. We do not know whether disease is the consequence of a primary flaw in their mode of *production* or because inadequate *consumption*—of antibodies complexed to antigens—is responsible. The studies of complement receptors were directed, in part, at the latter hypothesis.

Sitting in that secular temple of the English Enlightenment, we were far removed from the hospital wards of New York City. I wondered how soon it would be before the victories of molecular biology resulted in armistice at the bedside. I remembered Dolores, who braved madness and gangrene—and six stillbirths—to give birth to her only daughter. Then Sheila, whose disease waxed and waned with her menses and whose mind became muddled in the process; the prednisone which permitted her to manage a ballet company disfigured her profile. Agnes took ill with fits, thought to be due to high blood pressure, and her legs swelled to balloon size as her kidneys failed; after the drug cytoxan wiped out her bone marrow she made a tentative recovery and enrolled again in a master's program. Balanced between this dangerous disease and unforgiving drugs, these women and their doctors await news from the front. It is no great comfort to a desperately ill patient, or to the anguished house officer charged with her care, to be assured that the *general* outlook for patients with this disease has improved.

The threats to health in Restoration England did not include lupus. John Graunt and William Petty analyzed the bills of mortality of London, and found that by 1676 of "100 births, 36 died before the age of six, and only one lived to 76; one person in a thousand died of gout." Graunt worried that rickets was becoming the most common disease of the age. He was able to arrive at these earliest of vital statistics because the bills of mortality had become accurate enough to provide evidence for these conclusions. Their improved accuracy was in large measure a response of the government to a disaster of public health, the great plague of London in 1665. Of a population of almost half a million, seventy thousand died in *one summer*. Since the bubonic plague is attributed to the transmission of *Yersinia pestis* to humans from infected rats by flea bites, it is not surprising that the poor died in greater proportion than the rich. It is my impression that among patients with lupus, the poor also fare worse than the rich, since in few other diseases is the constant attention of a skilled internist so crucial to the outcome. As in the reign of Charles, so also in the reigns of Thatcher, Gorbachev, and Reagan: I know of no society in which those of rank and fortune fail to use these assets in the service of their own health.

These social ruminations extended themselves to the origins of the Royal Society. It cannot have been accidental that the revolt of the skeptics followed so closely the Cromwellian revolution. A generation had exhausted itself in the conflict between Cavalier and Roundhead; the earthquakes of civil war and regicide had been succeeded by tremors of the trade wars. It can, indeed, be argued that both the founding of the Royal Society and the seizure of New York from the Dutch (1664) were delayed expressions of the militant, commercial spirit of the Puritan revolution. Christopher Hill, in *The Century of Revolution 1603–1714*, traces the intellectual roots of the Royal Society to the adherents of Cromwell who established an "invisible college" for the discussion of experimental science. Pointing out that "science entered Oxford behind the parliamentary armies," he identified among this group the first-secretary-to-be of the Royal Society, John Wilkins, Cromwell's brother-in-law; Jonathan Goddard, who was a physician to Cromwell's armies in Ireland and Scotland; William Petty, who was the surveyor for these armies; and John Wallis, who was the cryptographer for Parliament. This nucleus of the invisible college was removed to London in the course of the Restoration, but not before it had sparked a new generation by the Isis: Wren, Boyle, Hooke, and Locke. Science then departed Oxford—some say indefinitely—for London and Cambridge. In 1669, Robert South used an official Oxford occasion to condemn "Cromwell, fanatics, the Royal Society, and the new philosophy." Many of the former Oxonians were physicians with a broad range of talent: Goddard not only developed his own drugs but became one of the first English makers of telescopes, William Petty was not only a biostatistician but also a surveyor, and John Locke came to medicine after undergraduate enthusiasm for mathematics and Oriental studies. (Locke also helped frame the constitution of the Carolinas.)

When the young Wren gave a speech on the new philosophy at Gresham College in 1660, the invisible college reformulated itself as a society, formally obtaining its charter two years later. David Ogg lists some of the subjects discussed in the first published *Philosophical Transactions* of the Royal Society of 1665:

> an account of improved optic glasses made in Rome; there was a communication by Hooke intimating that, with a 12-foot telescope, he had seen a spot in the belts of Jupiter; to this was added an account of a book in press—Boyle's *"Experimental History of Cold"*; there was also a description of a "very odd, monstrous calf born in Hampshire"; and finally there was a short article on the new "American whale fishing about the Bermudas." Other issues produced accounts of "The mercury mines of Friuli; a method of producing wind by falling water; revelations by the microscope of minute bodies on the edges of razors, on blighted leaves, on the beard of the wild oat, on sponges, hair, the scales of a sole, the sting of a bee, the feathers of a peacock, the feet of flies, and the teeth of sharks; a baroscope for measuring minute variations in the pressure of the air; a hygroscope for discerning the watery steam in the air [these were Hooke's inventions]; Mr. Wings's Almanac giving

the times of high water at London Bridge; a new way of curing diseases by transfusion of blood [!]; the process of tin-mining in Cornwall; and the making of wine in Devonshire."

In the *Philosophical Transactions*, "useful knowledge" took precedence over theory, in much the manner of the French *Encyclopédie*, which recapitulated the English example, but with better line drawings. The utilitarian aspect of the new Society, its openness to accidents of the natural world, to the new landscapes of telescope and microscope, reflected not only the commercial adventurism of Puritan England but also the imperial ambitions of the young monarch. Under Charles II, son of Cavalier and son of Roundhead joined in common cause to fight the Dutch for dominion of the sea.

John Dryden included an apostrophe to the Royal Society in his poem "*Annus Mirabilis*: The Year of Wonders, 1666." In this "historical" poem, Dryden contrived simultaneously to celebrate the English victory over the Dutch, free enterprise, and the distillation of useful knowledge from divine law (in the last stanza, a Limbeck is a distillation flask).

> *But what so long in vain, and yet unknown,*
> *By poor man-kind's benighted Wit is sought,*
> *Shall in this Age to* Britain *first be shewn,*
> *And hence be to admiring Nations taught.*

> *The Ebbs of Tides and their mysterious Flow*
> *We, as Art's Elements shall understand*
> *And as by Line upon the Ocean go*
> *Whose Paths shall be familiar as the Land.*

> *Instructed ships shall sail to quick Commerce*
> *By which remotest Regions are alli'd;*
> *Which makes one City of the Universe;*
> *Where some may gain, and all may be suppli'd.*

> *O truly Royal! who behold the Law*
> *And rule of Beings in your Makers mind:*
> *And thence, like Limbecks, rich Idea's draw,*
> *To fit the levell'd use of Human-kind.*

Certainly among the first to gain and to be supplied was the young monarch. Charles wrote his sister that "the thing which is nearest to the heart of the nation is trade and all that belongs to it." What "belonged to it" was the scientific foundation for mastery of the sea as the highway of commerce. God and His English Church were not exempt from this national endeavor; Bishop Thomas Sprat, in his *History of the Royal Society* (1667), asked, "If our Church should be an enemy to commerce, intelligence, discovery, navigation, or any sort of mechanics, how could it be fit for the present genius of this nation?"

It is difficult to know whether the national consensus as to what knowledge was *useful* determined that the Royal Society paid more attention to the telescope than to the microscope. Science is generally conducted both for curiosity and for utility and it is usually impossible to disentangle the two motives. But, in retrospect, it is possible to remark on a branch point of decision—conscious or otherwise—in the 1660s. London was decimated by the plague, water was unfit to drink, sewage flowed through the streets, and yet the small objects observed by the microscopes of Hooke were largely ignored. They remained playthings for amateur curiosity, while data obtained by means of telescope and barometer were transformed into the edifice of modern physics. The Fellows of the Royal Society fulfilled their goals: that natural knowledge "shall in this age to *Britain* first be shewn." Astronomy guaranteed navigation, optics guided landfall, and calculus let ballistics rule the waves.

Given the development of microscopy, there would seem to be no a priori reason why curiosity should not have driven the experimentalists of the Royal Society to develop biology as rapidly as they established physics. For reasons that remain unexamined, it was not until the nineteenth century that it became possible to establish that some of the objects observed by Hooke's microscope were the causative agents of disease. One suspects, however, that the Fellows of the Royal Society were not unmindful of the agenda of Empire. Had England exported wine, Hooke might have been a Pasteur.

These reflections suggest that it might be difficult, indeed, to decide how much we owe the establishment of the Royal Society to the Cavalier curiosity of the astronomer-king, and how much to Roundhead Oxford ("no knowledge but as it hath a tendency to use"). As Lawrence Stone has suggested in *The Causes of the English Revolution*, "To make sense of these events, to explain in a coherent way why things happened the way they did, has necessitated the construction of multiple helix chains of causation more complicated than those of DNA itself. The processes of society are more subtle than those of nature."

Other men's disciplines appear cleaner than one's own. DNA has turned out to be more than Watson and Crick bargained for, and the reasons why patients with lupus make antibodies to those two strands remain pretty much a puzzle. But Stone's evocation of DNA with respect to the Puritan revolution brings the seventeenth and twentieth centuries together in one other way. It is no accident that we have witnessed an explosion of molecular biology and immunology in the Western democracies since the Second World War. Government and industry have directed considerable proportions of our treasure to the basic sciences necessary for the understanding of disease, and sufficient money has been found so that we live under conditions "where some may gain, and [by and large] all may be suppli'd." We have gone to the moon, spliced genes, *and* cured rickets and gout. We've built enough weapons to destroy our enemies and ourselves many times over, and are about to apply Boyle's law to the revival of nerve gas, but we have also discovered drugs for childhood leukemia and cut remarkably the death rates from stroke, diabetes, and heart attacks. It is probably no accident either that commerce and trade have flowed from each of these activities. Meanwhile, the Soviet Union, filled with the most capable of scientists, has, like seventeenth-century England, directed its attentions almost exclusively to the physical sciences. Not unpredictably, the Soviet system has yet to discover a drug. Alone of advanced countries, Russia has experienced a decline in the life span of its citizens. Lamentably, one of the few Soviet scientists who might have made a contribution to our afternoon's discussion at the Royal Society is not permitted to travel abroad.

Persuaded that my amateur incursions into the territory of Professor Stone were likely to be considerably less productive than his reflections on DNA, I turned my attention back to the hall and immunology, where we were brought up to date on one of the major research accomplishments of the decade: isolation of the T-cell receptor for antigens. These receptor molecules on the surfaces of T-lymphocytes function as cell-bound analogues of circulating antibodies. Isolation of the T-cell receptor and the identification of its genes has put an understanding of cellular immunology within our grasp. Moreover, as Ellis Reinherz of Boston told us, these receptors occur in families of proteins which vary in their display on the surfaces of cells as the cells grow and develop. In consequence, by using fluorescent, monoclonal antibodies, one can spot clones of T-cells as they mature in the thymus, much as one can pinpoint hurricanes early in the course of their Caribbean origins.

This story of painstaking analysis and occasional legerdemain was followed by an equally impressive feat of scientific virtuosity, related by Jonathan Uhr of Dallas. Uhr and Ellen Vitetta, in experiments obviously as pertinent to cancer as to immunology, had joined the techniques of protein chemistry and immunology to fashion a kind of "magic bullet." By linking a plant toxin (ricin) to a monoclonal antibody which recognized an abnormal cell, his group had in hand the means of realizing the dream of Paul Ehrlich: killing unwanted cells in the thymus, for example, without injuring normal ones.

Putting the two presentations together, a strategy emerged. Discussion brought out that it should be possible to eliminate those abnormal cells—in Dolores, Sheila, or Agnes—which control the production of the self-directed antibodies. One last hurdle remained, a small obstacle compared to those already overcome: A way would have to be found to direct the hybrid toxin/antibody molecule toward the patient's aberrant T-cell clones; some method for that was surely being researched. Surely it was only a matter of time before all the knowledge we had witnessed here, all this skill, would be of use. Pride in what we had heard, pride in the crisp solution of ancient problems by the new techniques of molecular genetics, fueled the discussion. It was difficult not to feel somewhat self-congratulatory about this matter, especially here, pleased with the history and sense of place, pleased with the concordance of dazzling experiment and obvious social utility: the conquest of disease. The doctrine of *Nullius in Verba* had undoubtedly led to this point in our science; that which was demonstrable by experiment was not only true, but good.

After the meeting broke up in this mood of bonhomie, some of the visitors were invited upstairs into the library of the Royal Society, where a scholarly assistant exhibited a copy of the Signature Book of the Society. It is the custom for each Fellow to sign his or her name immediately upon election; the visitors were able to scan three centuries of eminence. From the most recent era, one could turn back through autographs which constituted the histories of biology, chemistry, and physics. Finally, back amid the calligraphy of unknown earls and dukes could be found the names of that earliest group of Fellows. There, more alive than in portraits, were the signatures of Wren, Newton, Boyle; to the bookish, the word is worth a thousand pictures.

These signatures evoked not the history of events, or of class, but of sentiments. What one sensed from these autographs was a kind of resonance of optimism, a resonance over that wide span of years between our young science of experimental medicine as revealed in the lecture theater and the young sciences of physics and chemistry in 1662. Boyle was thirty-five, the king was thirty-two, Wren was thirty, Sprat and Hooke were twenty-seven, Newton was twenty. The revolution of the skeptics, to which we can trace our science, was an achievement of youth; in a very real sense, we recapitulate that springtime each morning when we walk into the lab. *Nullius in Verba* is the scientist's equivalent of "Play ball!"; it might be said that experiments express the doubts of our perpetual adolescence. Sprat had it just right, spelling it out for us in the library of his Society, that we are "beholden to Experiments; which though they have not yet completed the discovery of the true world, yet they have already vanquished those wild inhabitants of the false world that us'd to astonish the minds of men."

17

SPRINGTIME FOR
PERNKOPF

Considering that the decade began in the midst of worldwide economic depression and ended in worldwide war, nothing short of a crack performance could persuade one that the 1930s were a time "When the Going Was Good." The performance by Evelyn Waugh in a travel book of that title is definitely of concert grade. In ravishing prose Waugh recounts his journeys to "the wild lands where man had deserted his post and the jungle was creeping back to its strongholds," lands which in that volume are represented by British Guiana, Egypt, and the borders of South Africa and Brazil. So light is his touch and so sure, so free from the shadows of that totalitarian decade, that we are placed at peril of nothing more wicked than inconvenience or dirt.

Here is Waugh describing an insect-ridden border town in the Brazilian jungle:

> Sunday Mass was the nearest thing to a pretty spectacle that Boa Vista afforded, and the men assembled in fair numbers to enjoy it. They did not come into the Church, for that is contrary to Brazilian etiquette, but they clustered in the porch, sauntering out occasionally to smoke a cigarette. The normal male costume of the town was a suit of artificial silk pyjamas, which many of the more elegant had washed weekly.

These are the cadences of pre-*Brideshead* Waugh, the battle sounds of Waugh on the ramparts of order. Here he is a bit later, fixing forever London's slovenly Bohemia with a description of his artsy hostess: "The hair, through which she spoke, was black."

Last summer I followed this Waugh of 1930 on his journey east of Suez, deserting—temporarily—my post on the jungle battlegrounds of Massachusetts, where wild grape lays siege to an embattled lawn. Waugh, by way of Djibouti, was en route to the coronation of Haile Selassie (alias Ras Tafari) as emperor of Abyssinia.

Waugh's itinerary, described in lavish detail, can make sense only to one whose grasp of East African geography is firmer than mine; it requires at least a cursory appreciation of why the direct railway from Djibouti passes through the termini of Diredawa and Hawash while bypassing the metropolis of Harar. To find out, I left my seaside chair to consult an old wall map we had recently acquired at our favorite junk shop in Falmouth. The map is of the traditional sort that used to hang in elementary-school classrooms from Bangor to Ventura. Six independent parchments are suspended in window-shade fashion from a common roller; when new, they flicked up at the teacher's tug. They display the several continents divided into a patchwork of sherbet-colored nationalities. Individual scrolls show North America, South America, Europe, Asia, Africa, and finally "The World on Mercator's Projection; by Cram." This last map is dominated by Britain's vast holdings, marked by traditional pink: Vancouver to Tasmania, India to South Africa, Guiana to Palestine. When I flipped the scrolls to Africa, Addis Ababa appeared in the middle of a lime-colored region called Italian East Africa, as the capital of "Ethiopia (formerly Abyssinia)." The lime region included Eritrea and Italian Somaliland, holdings which bracketed two bite-size pieces on the Horn, French and British Somaliland. Djibouti was the capital of the French colony and the major port nearest to Addis Ababa.

My curiosity was immediately satisfied on one point: Djibouti and Addis (as Waugh's reporter friends called the town) were on a straight line, whereas Harar was a bit off to the right. Diredawa and Hawash didn't merit the cartographer's attention. But my own attention became irretrievably diverted from Waugh's journey to another topic when I discovered that it was possible to date the drawing of this map to the spring or summer of 1938.

From the map of Africa alone, it was easy to establish broad time constraints: Clearly, the national boundaries had been drawn after late 1935. That was when Abyssinia had fallen into Mussolini's pocket after a cruel and confusing military campaign which Waugh described as a tragic musical comedy. Haile Selassie's European support had been fatally eroded by an agreement between the English and French foreign ministers, Hoare and Laval; the subsequent careers of these two civil servants did not establish their bona fides as staunch defenders of freedom. The map was also charted before the Second World War, during which Italy was stripped of her temporary empire, and considerably before postwar upheavals in Africa had fractured that continent's boundaries forever. Belgium, France, England, and Portugal were still shown in possession of much of Africa.

From the map of North America, unchanged in boundary or aspect, one could deduce only that it had been drawn before air-conditioning tilted the demographic scale. Appropriate symbols indicated that the populations of Miami and Phoenix were less than 50,000; Tampa and San Diego each had between 50,000 and 100,000 inhabitants. The map antedated wartime shifts of populations; Los Angeles had between 250,000 and 1 million inhabitants.

But it was the map of Europe that made it possible to date the moment of its drafting to the spring or summer of 1938. For there, smack in the center of the continent, squatted a beige Germany, not only undivided into East and West but in firm possession of beige Austria, from which no boundary line separated it. The drafters of the map had not troubled to indicate that the area surrounding Wien (Vienna) was "formerly Austria" in the manner of Abyssinia. Thus the map was clearly formulated after the Anschluss of March 11, 1938. Adolf Hitler had preempted an Austrian plebiscite to annex an eager Germanic population to his Reich. Since Czechoslovak territory remained distinctly outlined in lemon-yellow, retaining its German-speaking regions (the Sudetenland), the map had obviously been framed before the Munich crisis of September 1938. That ill-fated attempt at "peace in our time" on the part of England's Chamberlain and France's Daladier ceded the Sudetenland (and the bulk of Czech ground fortifications) to Hitler in return for unkept promises. Churchill scolded Chamberlain in Parliament: "You were given a choice between dishonor and war. You have chosen dishonor and you shall have war!"

So there was the Europe of 1938, hanging on the wall of our summer cottage like a three-year-old calendar. In 1938 the bulk of Europe's people were ruled, not unwillingly in the main, by regimes that were autocratic, if not overtly totalitarian. Poland and Hungary were under control of the sort of right-wing juntas or oligarchies that nowadays disfigure Latin America. Germany had Hitler, Italy had Mussolini, and Spain was in the process of acquiring Franco, whose armies had just succeeded in splitting the Republican forces into two embattled halves. In the Western democracies, anticommunists competed with the self-interested for the right to avoid the news from Central Europe and Spain. On the other side of the globe, Japan's military cabal sat astride Korea and Manchuria; its bombers were strafing civilians in China. The going was not good for the parties of reason.

But the going was clearly good that spring for some people, among them Professor Dr. Edward Pernkopf. I stumbled upon his career as I followed the map of 1938 into the medical literature. Pernkopf, an anatomist and embryologist, was Komissarischer Dekan ("official dean") of the medical faculty of the University of Vienna. The dean's public shenanigans remain available for inspection today in bound volumes of the *Wiener Klinische Wochenschrift* ("Viennese Medical Weekly," or *WKW*, as we'll call it). Organ of the Viennese Medical Society, its pages were filled with the accomplishments of the School of Vienna: Wenckebach and Molisch, Schick and von Pirquet, Boas and Pick. By 1938 the journal enjoyed a reputation by no means inferior to that of today's *New England Journal of Medicine*.

Pernkopf's name pops up on the masthead for the first time in the issue of May 20. His introduction is remarkable in the way that it describes the auspices of the journal: "published by members of the medical faculty in Vienna as *represented by* [my italics; the German is *vertreten durch*] Professor Dr. E. Pernkopf, dean." In slightly smaller type are listed the executive editors: Professor Dr. M. Eppinger and Dr. E. Rizak. When we turn to the immediately preceding issue of the *WKW*, that of May 13, we are presented with a different cast of players. On the cover of that issue, the journal simply describes itself as "published by members of the medical faculty in Vienna." Presumably something had taken place that required the faculty to be "represented." The editorship on May 13 was in the hands of four distinguished academicians, most notably Professor Dr. H. Chiari, who is best known to us for describing thrombosis of the hepatic veins. Both the May 13 and May 20 issues are part of volume 51. What produced this sudden change in the leadership of the *WKW*? Editorial boards usually turn over as volumes change in July or January. Why was the faculty now "represented" on the journal by their dean? How did Eppinger and Rizak displace Chiari and company in the middle of the year?

Back to the map of Europe. Hitler's troops arrived on March 12, to the delight of multitudes and the agony of a few. Newsreels and photos of the time show the adoring crowds that greeted Hitler and his entourage. The beatific expressions remind one of those seen in our decade on the faces of the young at a rock festival or their elders between halves of the Superbowl. But Hitler had to convince the world that the Anschluss would be as popular over the long haul as it was in the instant. So, for three weeks he waged an electoral campaign for a vote of yes in a plebiscite on the question of union between Germany and Austria. The campaign was carried out on both sides of a still-recognized border; only after over 99 percent of eligible Aryans had voted in favor of the Anschluss on April 10 were the maps changed. (My old school map must have been drawn up after April 10.) One month later Pernkopf was perched on top not only of his faculty and their journal, but of the party hierarchy. The Nazis had begun to tighten their grip on what was now no longer called Austria, but had become "the Ostmark."

In the first issue over which they exerted direction, Pernkopf, Eppinger, and Rizak inserted a special page, just inside the cover, on which they charted the party line to come. They describe their own reaction to the advent of Adolf Hitler as a *Begeisterungssturm* ("gale of rapture"). In the same spirit of Teutonic understatement, they swear undying allegiance to the new Reich, promise to make the *WKW* an organ of wisdom to serve the Fatherland, and pledge that they will transform "Vienna, the oldest University in the Reich, [into] the stronghold of the new German world." But the reader soon learns that joy, rapture, and patriotism are not the whole story. For in that fateful issue of May 20, we also find printed the first official speech of Dean Pernkopf to the faculty and students of Vienna. Entitled "National Socialism and Science," it had been read from the venerable lectern of the anatomical theater; it, too, is a spectacular performance:

> What was the dream of our youth, what we dared not hope for, has become reality: we are one people, one Reich, one leader [*Ein Volk, Ein Reich, Ein Führer*], Adolph Hitler; he strides before us and we follow him gladly.

Since the speech was given on April 6 (four days before the plebiscite) by an official of the government, one gets the notion that the voting wasn't expected to produce a cliffhanger. The dean expanded further on the new order:

> Here you will be educated as doctors, that is as German, as National Socialist doctors. . . . I believe it is entirely in order to explain to you how deeply the Idea of National Socialism must permeate our education and our science in order for us to arrive at our goal. But you may well begin by asking me: what, in fact, does National Socialism have to do with science at all? And I can only answer you in the following manner: National Socialism is not just a bare idea, not a bare theory in

the service of a politically motivated mobilization of strength; for us it is more than that: it is a view of the world [*Weltanschauung*] and as such every expression of our spiritual life, all our will and striving, thinking and acting. National Socialism permeates and fertilizes those not only in a general sense, but also in each particular.

Now, rhetorical flights of this sort seem to constitute a kind of regional academic dialect in German; it seems clear, to me at any rate, that prolonged exposure to the works of Hegel, Schopenhauer, and Nietzsche may tune the mind, but dull the ear. Pernkopf cuts the academic rhetoric soon enough to come down to brass tacks:

> Some will say, yes, but our science must be free, free from every external and internal influence. Freedom, my comrades, in the liberal sense which leads to chaos, this freedom which foreign powers parade on their flags—but really only wish to exploit—this sort of freedom cannot and will not be permitted for science. Here we must have direction, planning, and goals. Here we will have planned order [*Ordnung*] in science, exactly as in art and in economics. The idea of art for art's sake which has set heads spinning in our time has its parallels in science as well: for many it's been science for the sake of doing science; indeed for many it was the only sense of their work. I must tell these people quite openly: Such a conception of our intellectual life strikes me actually not as a selfless endeavor, but rather the expression of a self-seeking narcissism of intellectual underachievers who wish to hide their vanity in the cloak of altruism.

The dean seems to have anticipated some of the rhetoric of our own recent past (the phrase "nattering nabobs of negativism" springs to mind). But his anger takes a nastier turn when he continues to belabor the theme that the only useful goal of art and science is service to the nation, to the *Volk*. Pernkopf ties this notion to a warning against foreign influences, sounding notes that seem to come from Bayreuth:

> Think of how a foreign spirit—which unfortunately was disseminated from Vienna—tried to disrupt our music by promoting dissonant chords, how the atonal direction of melody—as the musical expression of Will and Idea in the sense of Schopenhauer—threatened to destroy our beautiful German music: who would deny the foreign origins of these corrupting trends? Indeed, proof that these trends owed their power to alien influences is afforded by the observation that when— thank God!—these influences were rendered powerless, they sank without a trace.

Translation: Mahler, Berg, and Schönberg ruined German music. Many of the prominent composers of dissonant or atonal music were not only cosmopolitan but Jewish. When Austrian anti-Semitism and pressure by National Socialists succeeded in removing Jews from prominent positions in the musical world, the influence of atonal music waned; Vienna could return to three-quarter time.

As in art, so in science. In passages of prose as clotted as Schopenhauer, our dean warns his audience that the same poisonous influences are at work in natural science and philosophy. The result, he moans, is a progressive leveling, a reductionism that muddies the clear waters of German reason. He turns to his own science, developmental anatomy. (Professor of anatomy since 1933, Pernkopf had edited a detailed atlas and published work on descriptive embryology.) Since National Socialism is "devoted to the practical solution of problems," he lists two critical issues that anatomy and embryology can address:

> Two concepts derive from our understanding [of the problems of human development] which particularly confront us with respect to our National Socialist *Weltanschauung*: the concept of [innate] constitution and the concept of race.

Pernkopf details how constitution and race are intertwined; some races seem to have within them the capacity to strengthen the constitution of an individual, whereas others provide "race markers" (*Rassenmerkmale*) that weaken the constitution. The dean promises his students that in their new German state all the disciplines of the medical faculty—not anatomy alone—will help them to understand the problem of race. The new curriculum will include race physiology, race psychology, and race pathology.

But the keystone will be genetics—the methodology of proband analysis, family and population studies, and the study of twins:

> That which was studied before for reasons of pure science or research, can now be of use in the fields of sports and occupational medicine, in marriage counseling, and in the determination of racial origins or paternity.

The dean summarizes the role of medicine in the new state:

> To assume the medical care—with all your professional skill—of the Body of the People [*Volkskörper*] which has been entrusted to you, not

only in the positive sense of furthering the propagation of the fit, but also in the negative sense of eliminating the unfit and defective. The methods by which racial hygiene proceeds are well known to you: control of marriage; propagation of the genetically fit whose genetic, biologic constitution promises healthy descendants; discouragement of breeding by individuals who do not belong together properly, whose races clash; finally, the exclusion [*Ausschaltung*] of the genetically inferior from future generations by sterilization and other means.

The peroration of this social Darwinist is appropriately laced with a quote from the most forceful advocate of that creed:

As Adolph Hitler said, a state that bases its *Weltanschauung* on biological thought, a state which in this time of racial pollution dedicates itself to its best racial elements, which in this time of an aging population and declining birth rate wishes to return to its earliest vigor, will set a task for the doctor not limited to his profession but extending to the life of the people. . . . We thank him [Adolf Hitler], the prophet of National Socialist Thought and the new *Weltanschauung* in which the myth of blood and the heroic spirit have been woken again, he, the greatest son of our homeland, and we wish to tell him at the same time that we, as doctors, will gladly place at his service our lives and souls. So shall our cry express what we wish with all our hearts: Adolph Hitler, Sieg Heil, Sieg Heil, Sieg Heil!

Several weeks pass. And, in an issue of the *WKW* devoted to an introduction by Pernkopf of his faculty before the annual postgraduate course, he again expresses the hope that from Vienna will shine the light of racial hygiene, under the heartwarming spirit of the Führer.

The dean thanks profusely the director of the course (our other friend from the *WKW* masthead, Dr. Rizak) and turns the podium over to the first lecturer. You will not be surprised to hear that it is Professor Dr. Eppinger.

The trio appears a few weeks later as joint signatories to yet another special dedication, in an issue of the *WKW* dedicated to the German Society of Physicians and Experimental Biologists. The dedication reads: "The Ostmark has returned to the Motherland!"

After the usual avowals of allegiance and a self-serving announcement of how the *WKW* will spread the news of science to the East, the long paragraph ends with the assurance (signed by Eppinger, Pernkopf, and Rizak) that "like every citizen of our state, the physician of the Ostmark wishes to be nothing more than a humble assistant in the work of our leader, Adolph Hitler."

Judged by their self-advertisements in volume 51 of 1938, our triumvirate may not have been humble, but they certainly placed the *WKW* in service of their leader. In keeping with the medical marching orders of the Third Reich, they print a remarkable document by an SS-Obersturmführer, Dr. A. Rollender. The author's academic affiliation is listed as the "Evening Study Sessions of the Physicians of the SS-Superdivision '*Donau*.'" You will not be astonished, in the *WKW* of Pernkopf *et al.*, to find the dean's favorite words in the title, "Race Biology as the Guideline of our *Weltanschauung*."

Rollender takes for his text the words of the Nazi philanthrope Alfred Rosenberg: "Belief in the worth of blood and the worth of the German race is the basic tenet of the National Socialist *Weltanschauung*." Rollender expands on this theme, sprinkling *Weltanschauung* as frequently about his paragraphs—and with the same rhetorical effect—as present-day contributors to *New York* magazine pepper their text with "life-style." He argues that a hot war rages between two *Weltanschauungen*. In the old, discredited liberal societies, the needs of the individual are favored over those of the nation-state. Those societies regard the nation (*das Volk*) simply as a collection of individuals. In consequence, liberal and Marxist [sic!] societies have no concept of heroism or self-sacrifice, which are grounded in blood relationships as reified in the state. In contrast, the new Germany is united by common blood, shared genes, and a heritable culture.

Biology, as interpreted by Rollender, plays the role of dealer in a Calvinist game of blackjack. The genetic makeup of an individual sets the limits of human free will. Science documents the fundamental inequality of men, and the scientific discoveries of our time have given us the basis for the deepest insights, which frame our *Weltanschauung*:

> National Socialism, which finds its meaning in the denial of individualism, in its opposition to formerly all-dominant liberal circles, found help for its political policies, its *Weltanschauung*, in the discoveries of Nature which, at the turn of the century, led to the rediscovery of Mendelian laws and to a series of related scientific disciplines which have reached considerable heights.

These "disciplines" are the concern of Pernkopf and friends: racial hygiene, "constitution" studies, the genetics of fitness. Braced by this gene-babble, Rollender argues against the tendency to equality (*Gleichheitstendenzen*) and for the new order of a racially determined (*rassenbestimmten*) society.

Now, all this puff and prattle, this posturing over blood, myth, racial purity, and so on might strike the reader of today as one of those cranky monologues spouted by a Hollywood Nazi of the 1940s as he trains his gun on Humphrey Bogart or Ray Milland. But the humor of these *Volk*, their *Weltanschauung*, their *Rassenbestimmung*, is gallows humor. "Exclusion of the genetically inferior from future generations by sterilization and other means" was the beginning of the long road to Auschwitz. The "study of twins" was the overt purpose of Mengele's pick and led to Sophie's choice. The confusion between animal husbandry and public administration led to the docket at Nuremberg.

No explanation of the behavior of Pernkopf and his friends is satisfactory. They were "evil" (à la Hannah Arendt) but no worse than Heidegger or Heisenberg, and they were by no means "banal." Pernkopf, Eppinger, and Rizak were eminent Viennese academicians. Cultivated in art and science, they could distinguish Hindemith from Schönberg after ten bars, or lues from cirrhosis at ten paces. They were purposeful men who knew the kind of society they wanted for their country and found a politics that would give it to them. It seems to me that these men were not suddenly seized by mass paranoia: Their personality did not suddenly "double" (à la Robert Jay Lifton). No, in the center of Europe, mistaken men, certain and arrogant, planned a broad racial experiment, told their students—and the world—what they planned to do, and went about doing it.

After rummaging through this melancholy record in the Viennese literature (courtesy of the remarkable library of the Marine Biological Laboratory), I wondered what the doctors in the Western democracies were reading about Pernkopf and company. The *Journal of the American Medical Association* is almost mute in 1938 on the issues of National Socialism in Germany and Austria. However, in November there appears a telling account entitled "The Fate of Austrian Scientists" from the Berlin correspondent of the *Journal*. The letter is dated September 26, 1938, and also details experiments by A. Hoffman on "The Influence of Training on Skeletal Muscles" and an appreciation for the eightieth birthday of Friedrich von Müller, a Munich internist.

Our correspondent notes that about half the assistant professors or instructors holding office at the time of the Anschluss have lost their positions. "The Jewish element has been prominent among these groups, whereas but few Jews have served as full professors in recent years." The correspondent continues with his depressing tale. Egon Ranzi, professor of surgery, was dismissed from his position as chief of the university clinic because of his support of Kurt von Schuschnigg (the ousted Austrian chancellor). Professors Arzt and Kerl, "Aryan" dermatologists, met the same fate as Ranzi and for like reasons. Indeed, our Berlin-based informant indignantly writes:

> Professor Arzt, decided anti-Semite, ardent proclerical and nephew of a late archbishop of Vienna, was in custody for a short time. Ernst P. Pick, professor of pharmacology, was forced to retire on account of being a Jew. As is generally known, Prof. Otto Loewi, who not long ago shared a Nobel prize with Sir Henry Dale of London, has been deprived of his post and spent some time in custody. . . . Foremost among Viennese psychiatrists and neurologists to be affected was Sigmund Freud. . . . Among the internists who have lost their positions are: G. Hitzenberger, radiologist; David Scherf, cardiologist; Julius Bauer; Karl Glaessner and Walter Zweig.

The last named was a pupil of Ismar Boas. That aged gastroenterologist had made Austria his refuge from Nazi Germany (and continued his active laboratory investigations on bile pigment). After the annexation of Austria, he ended his life with an overdose of barbital. Other suicides were reported: the pediatrician Professor W. Knoepfelmacher; Professor Oskar Frankl, gynecologist; and the dermatologist Gabor Nobl. Our correspondent concludes this list—which I have truncated—with the hope that it provides some "idea how the change in the political status of Austria has affected the faculties of medicine."

By September, then, the first effects of Pernkopf's *Ausschaltung* were obvious. By September 30 (the day of the Munich compromise, and the last day on which my map could have featured an independent Czechoslovakia), all medical men of Jewish birth or faith were excluded from the profession. Exception was made only where the number of Jews in a region was considered excessive for Aryan practitioners. For that reason, 357 men and 10 women were permitted to practice among the 140,000 Jews of Vienna. They could retain the title of "doctor" but were stripped of any academic rank. A placard bearing a blue Star of David in a yellow circle was required on the office door; similar devices were required on prescription pads and letterheads. The Jewish doctors could not treat or provide consultation for non-Jews; the heavy penalties included deportation. Jeering crowds of their former neighbors, supported by storm troopers, frequently formed vigils during office hours, forcing some Jewish patients to scrub the sidewalks of the city on their knees.

Emigration was difficult that spring and summer but not completely impossible. The daily routine of the suddenly unemployed began with waits on endless lines at the consulates for visas: exit or transit visas for the lucky, residence visas for the blessed. Any harbor seemed safe in that storm; visas were gratefully accepted to places unknown to the refugees except by name: Shanghai and Santiago, Rio and Durban, Oakland and Melbourne, Havana and Portland.

Almost fifty years later, the record of that closing trap has been almost completely erased. Many of us look back to the 1930s as a chic time when the music, the clothes, the decor—the going—was good. The springtime of Pernkopf has not surfaced in serious literature, although Anthony Powell's Widmerpool may provide a fictional template for the Austrian dean. No, for me that period is evoked best in a medical journal, the voice of which has echoes of Waugh. On turning the dreadful pages of the *Wiener Klinische Wochenschrift*, I wondered whether the events on the Continent provoked any reaction in the English medical press. Was the change in the map the first change in the weather of Empire? The issues of *The Lancet* in 1938 yielded more drama than the usual run of fiction or travel literature. Here is a letter dated March 26:

> OUR COLLEAGUES IN AUSTRIA
> In view of recent events, we . . . express our alarm at the possible fate of our colleagues either on account of their medical or social views, or on account of their belonging to the Jewish race. We beg our colleagues in all countries . . . to do all in their power, whether by public protest, by public or private assistance, to stand by any member of our profession who may suffer hardship under the new regime.

The letter bore eighteen signatures, most prominent among which were Sir W. Russel Brain and Lord Horder.

This generous and openhearted appeal did not go unanswered. On April 2, Dr. Aubrey Goodwin expressed the concern of "the British medical profession [over] the possibility of a further accession of medical refugees from Central Europe." Goodwin pointed to the "large number" of German refugee doctors already in practice in England. These he considered at an unfair advantage with respect to native practitioners by virtue of "the distinction of being 'continental' practitioners, is in the eyes of the British public a mark of distinction. . . ." The very next week's mail (April 9) brings support for Goodwin's view from Dr. Frederick C. Endean, who is convinced that a "further accession to the British medical profession of medical refugees from Central Europe . . . will undoubtedly result in undue competition."

From the same mailbag, a letter by the eminent Samson Wright points out that "Goodwin's comments, however, show the difficulties that will have to be overcome before goodwill [the March 26 appeal] can be converted into constructive help." He documents that since 1933, when Hitler came to power, a total of only 187 German doctors had been permitted to settle and follow their profession; there were, at the time, over 50,000 names on the medical register. He suggests that whereas no one would welcome the unlimited entry of *all* refugees, the possibility should be left open to permit "certain carefully chosen individuals to continue their work here to the advantage of medical science and for the relief of human suffering."

Dr. Mary T. Day, in the last letter of the issue, asks all members of the profession who would like to join the signatories of the March 26 letter to send their names to the honorable secretary of the Medical Peace Campaign at 39 Southgrove, London N.6. Right beneath Dr. Day's altruistic letter a small news item had been inserted. Only a reader insensitive to the editorial irony of Albion could fail to remark on the situation of this note:

TASMANIA AND IMMIGRANT DOCTORS
The Tasmanian Parliament has passed a Bill amending the Medical Act in view of a possible influx of doctors. Fourteen applications from German medical men have so far been made. The amending Bill requires candidates for registration to be British subjects.

On April 23, *The Lancet* takes up the issue of immigration in its lead editorial, entitled "An Overcrowded Profession?" Admitting that the influx of refugee physicians is infinitesimal (the editors accept Samson Wright's figures of 187 out of 50,000), they nevertheless worry that too many of the 187 are concentrated in one overcrowded area, the posh consulting rooms of Harley Street. They complain that

all [the refugees] have profited by the belief of the public in the superior merits of foreign doctors—a belief by no means diminished by references, in newspapers and elsewhere, to the distinction of the scientists who have been forced out of Germany. When a man without English hospital connexions secures within a year a practice that appears as large as those of many consultants long known and much esteemed, it is only natural [to] wonder whether his success is wholly based on merit. To speak plainly, the prosperity so speedily attained by some refugees has done more than anything else to weaken the desire to help refugees as a class.

This mean-spirited mood soon gives way to more conventional, generous grumbles:

> The alternative to such help, however, is a further degradation of English tradition. Already our former reputation for generosity to those who are in trouble because of conscience or race has passed to the French.

The editors conclude that probably some, but certainly not many, Austrian physicians might be welcome. In June, perhaps softened by the season, *The Lancet* opens its arms to one of these. In a lead article, we read:

> It would be ungrateful to allow Prof. Sigmund Freud to make our country his future home without bidding him welcome. His teachings have in their time aroused controversy more acute and antagonism more bitter than any since the days of Darwin. . . . This is not the time to appraise his contribution, but to greet him in the hope that he may find peace and some joy among his many friends and admirers in London.

This conformance with the English tradition of "generosity to those who are in trouble because of conscience or race" had not persuaded all factions of English medical opinion. A letter signed "MD, MRCP," of April 30, complains:

> As one of those mentioned in your lead article who refuses so ungraciously to hand over his living to foreign refugees . . . I suggest that it would be better to send foreign refugees to the countries with large populations and few doctors, such as India, rather than admit them to overcrowded England. Can you possibly deny the correctness of this? Although it is admitted that the possibility of their earning a large income is not so great in a poor country like India, and the Hebrew has less standing in an Oriental country.

But overt bigotry of this sort appears rarely. The better nature of the editors surfaces in a lead article of May 28 entitled "New Tests of Scientific Truth," in which the editorialist of *The Lancet* refers to two Nazi articles, published in Germany and Austria, respectively. The German article, by a Dr. Karl Haberman in the *Münchener Medizinische Wochenschrift* of May 6, proposes to rid the field of those studies of psychopathology that have been contributed by Jews. The editorialist points out that inasmuch as Germans are now taught that their misfortunes have been caused by aliens in their midst, and that the Nordic outlook is right, it becomes natural that all influences not of their own racial origin must be defective and misleading.

The Lancet pokes fun at Haberman by the simple expedient of translating his views into English, leaving in place the necessary references to a new *Weltanschauung*. The writer comments that "we look forward with interest, though with some concern, to the International Congress of Psychotherapy when some of our German colleagues will meet us in the serene air of Oxford."

The second part of *The Lancet's* editorial is devoted to a summary of Pernkopf's speech in the May 20 issue of the *WKW*. Translation of the dean's florid German into simple English suffices as a method of ridicule. The editorialist regrets that two formerly distinguished medical journals, those of Munich and of Vienna, have succumbed to a new criterion of truth: subservience to a political regime.

> In raising ... aids to self-sufficiency, nationalism has now raised cultural barriers as well. Hence we have become familiar again with the purging of libraries and with bonfires of books.

The action in *The Lancet*, during late spring and summer, shifted to reports of Parliament. On July 14, the home secretary in Neville Chamberlain's cabinet, Sir Samuel Hoare (the Hoare who teamed up with Laval to sell out Haile Selassie), was pressed by members of the House:

MR. MATHERS: Has the right honorable gentleman received intimation of the concern of doctors practising in this country about the numbers [of refugees] who have already been admitted to practice here?

SIR S. HOARE: I have had discussions with representatives of the principal organizations, and we both agree that discrimination must be exercised. I think it will be found that we shall be able to admit a limited number and at the same time maintain effective discrimination.

MR. MANDER: Do we not want to take as generous an attitude as we can on this matter in accordance with our long cultural tradition?

No further answer was given.

I hear in that last sentence the cadences of Waugh ("The hair, through which she spoke, was black"). They differ appropriately from the Wagnerian brass of Pernkopf, Eppinger, and Rizak. Perhaps the most reassuring aspect of *The Lancet* is how little it has changed over the years. It is no less reassuring that today's *Wiener Klinische Wochenschrift* reads much like *The Lancet* of 1938. To paraphrase Pernkopf, proof of the power of National Socialism is the fact that when its force was removed, its effects on medical literature disappeared without a trace. In the index of the *WKW* for 1984, there is no entry for "race hygiene," any more than in *The Lancet* (or the *Journal of the American Medical Association*) of 1938.

One often hears, these days, from the partisans of this cause or that purpose, that the Western temper of our time is fatally flawed by its lack of political faith. Surveying the map of 1938 and reading the sorry history of my parents' generation, I am not displeased that the "discredited liberal democracies" continue to muddle along in secular disarray. Besieged from within and without by the ideologues of certainty—those for Christ, Islam, Talmud, or Marx—the posts have been manned by the soldiers of doubt. It is not the armies of doubt who have slaughtered Armenians, Jews, Biafrans, Cambodians, and black South Africans. The commissars and ayatollahs have not inscribed "art for art's sake" or "science for the sake of science" on their banners.

Pernkopf is the culprit, that spirit of arrogance in spring. Imagine what might have been, had the dean of the medical faculty of Vienna in 1938 been Bergebedgian, that "Armenian of rare character" whom Waugh met in the Abyssinian town of Harar. That refugee from the first holocaust of our century

> spoke a queer kind of French with remarkable volubility, and I found great delight in all his opinions; I do not think I have ever met a more tolerant man; he had no prejudices or scruples of race, creed, or morals of any kind whatever; there were in his mind none of those opaque patches of principle; it was a single translucent pool of placid doubt; whatever splashes of precept had disturbed its surface from time to time had left no ripple.

We need that *Weltanschauung* today.

18

THE BARON OF
BELLEVUE

I soon paid a visit to my friends, and related these adventures.
Amazement stood in every countenance . . . every person paying the
highest compliments to my courage and veracity.

R. E. RASPE, *Baron Münchhausen*

The medical resident had a big grin on his face as he came into my office
with his team of interns and medical students. He was bursting with news
of last night's admissions.

"You won't believe what came in—and on the prison ward, of all places.
The case is right up your alley. This guy doesn't have AIDS or hepatitis, he's
not even an addict. He's got Bartter's syndrome. What's more, he's been to
MIT, so you can get a history."

Our house staff at Bellevue has become a bit bruised from its battles
with the diseases of drug abuse. They are chronically reminded that hepatitis, endocarditis, and AIDS are preventable diseases in the population of
addicts. The toll is greatest on the prison ward, where a clientele of truculent
heroin users submits unwillingly to medical care. Communications between doctor and patient are jammed by the noise of race, language, and
rearing; most of our patients have not been educated on the banks of the
Charles. Moreover, they present themselves with rather uniform signs of
their afflictions: the jaundice of hepatitis, the fever of endocarditis, the
wasting of AIDS. Care of these unfortunates requires bonds of sympathy
rather than feats of reason. No wonder, then, that the resident was so excited
by the unusual case of Mr. Malone, the MIT computernik with symptoms
of a tricky disease. Bartter's syndrome is a rare endocrine disorder in which
the serum level of potassium is so low as to produce muscle weakness,

fainting, and paralysis. The low levels of potassium are accompanied by high levels of an adrenal hormone called aldosterone, and it is the determination of these two substances in the serum that alerts an astute doctor to the diagnosis. The syndrome was first described in 1962 by Frederic Bartter, then at the National Institutes of Health, and by now several hundred patients have been described. The condition must be distinguished from more common causes of low potassium and high aldosterone, such as overuse of laxatives, too frequent diuretics, or self-induced vomiting. Patients with Bartter's syndrome excrete greater than normal quantities of prostaglandins in their urine and that is why Mr. Malone's case was right up my alley. I've been studying prostaglandins, those hormonelike regulators of the circulation, for almost two decades.

Conference facilities were not uppermost in the minds of the builders of Bellevue's new prison ward on the nineteenth floor, so we listened to Mr. Malone's history in my office three floors below. Over coffee, the medical resident—we'll call him Mike—gave us the details. He consulted the medical record frequently, for Mr. Malone had spelled out copious details of his many encounters with physicians.

Mr. Malone was a burly giant of a man in his mid-forties. He was admitted from the Men's House of Detention, where he had been remanded for extradition to California on a charge of embezzlement. Immediately on his arrival in prison he had complained of severe weakness; the mention of Bartter's syndrome brought him to Bellevue.

His curriculum vitae differed in all aspects from those usually obtained on the nineteenth floor. He told Mike that he was a native of Louisville, where both of his parents were "high-ranking corporate executives." After prepping at Phillips Andover, he declined admission to Harvard and went to MIT because of his intense interest in computers. He majored both in electronic engineering and in computer science. He was in fine physical condition while in college, playing "intramural sports and making the MIT rowing team." Having graduated near the top of his class, he was snapped up by one of the emergent microchip hatcheries in the shoals of Palo Alto. For a decade or so he bounced on the waves of the American computer revolution, emerging with a brace of patents, various vice presidencies, a Porsche, and a perforated nasal septum.

Mr. Malone was vague as to how he had come to New York, or how he had fallen afoul of the legal system; he confessed to an excessive fondness for extracts of coca leaves and to the impact of this habit on the rerouting of company funds. He was much more exact with respect to his endocrine disease. For the past ten years or so he had consulted some of the more respected internists and endocrinologists in the Bay Area for what was undoubtedly a bad case of Bartter's syndrome. Mr. Malone told us that his serum ranged from 2.0 to 2.5 milliequivalents per deciliter (the normal range hovers around 4, and levels as low as those reported by Mr. Malone invariably produce symptoms). Happily, his disease abated somewhat when he munched on "a carload full of potassium pills," but he could never quite tell when he would become weak and faint-headed. At times of corporate crisis he was not infrequently confused.

He sought out the best authorities; they told him that his serum levels of aldosterone were "off the wall" and that his prostaglandin excretion was also high (an authentic point of diagnosis to Mike). Finally in 1980 or so he met Bartter himself at the "University of Texas Southwestern Medical Center in Dallas." He was advised to take a lot of potassium pills.

At about this point in the story, it was Mike's turn to present to us the results of the physical examination and the initial laboratory values. As he did this, he handed me Mr. Malone's chart and it confirmed what we had heard. Aside from muscle weakness, neither Mike nor the intern had found anything wrong on physical examination, nor were there any abnormal numbers on the laboratory sheet; the potassium was right where it should have been, at 4.1. Mike attributed this salutary state to the many potassium pills Mr. Malone had swallowed before admission.

Certainty in diagnosis is rare. But every once in a while the clinician is seized by a hunch so powerful as to compare with the hypotheses of laboratory science. I was sure that Mr. Malone was fibbing, that he was an outright liar, and that he no more suffered from Bartter's syndrome than I did. In the first place, nobody who ever rowed in Cambridge, Massachusetts, was on a "rowing team": they were on "crew." And second, Bartter hadn't left the NIH for Dallas, but for San Antonio. But even if those two slips were just honest mistakes, it would be too much of a coincidence to have a convicted, coke-sniffing embezzler come to Bellevue with the one syndrome that can be reproduced at will simply by taking laxatives or diuretics. I was sure that Mr. Malone did not have Bartter's syndrome; he had a classic case of the Münchhausen syndrome.

> Here is described a common syndrome which most doctors have seen, but about which little has been written. Like the famous Baron von Münchhausen, the persons affected have always travelled widely; and

their stories, like those attributed to him, are both dramatic and un-
truthful. Accordingly, the syndrome is respectfully dedicated to the
baron, and named after him.

So wrote Richard Asher in *The Lancet* in 1951, linking forever the name
of the baron with those patients who for unknown reasons invent their
medical histories. Asher pointed out that patients with this syndrome
shared four telltale signs. Their immediate history was acute and harrowing, but not entirely convincing; they responded to questions with a mixture
of truculence and evasiveness; they were likely to bear scars of random
surgery or reports of arcane diagnoses; and they invariably carried wallets
or handbags stuffed with hospital attendance cards, insurance claim forms,
and litigious correspondence. Such patients are rare, indeed, but not so rare
as to escape description and analysis in the clinical literature. Each year the
Index Medicus lists over a score of articles which recount the behavioral
extremes to which the quest for disease has driven these bearers of false
medical witness.

The motives that drive patients to simulate disease vary. While the
quest for drugs or shelter seems to motivate a good number, others seem to
want the attention of concerned young doctors and nurses. Still others seem
to take pleasure from the repeated invasions of the bloodstream that
accompany modern medicine. We should not judge these people too
harshly; a society that encourages the couture of Claude Montana or the
gestures of Twisted Sister should not be surprised that some of its kooks are
in bondage to rubber tourniquets.

Even children are not immune. In 1977, Roy Meadow described in *The
Lancet* two cases of "Münchhausen Syndrome by Proxy," which he defined
as the production of *factitious* disease in a child by its parent. He recounted
the baleful stories of two young children who suffered from complicated
maladies: a six-year-old whose urine was dosed with pus and sugar, and a
toddler whose mother gave it so much salt by stomach tube that the child
eventually died. Münchhausen-by-proxy accounts appear yearly in the
pediatric literature.

The baron's legacy can be propagated. C. M. Verity *et al.* reported in the
British Medical Journal in 1979 a child with what they called Polle's syndrome, which they defined as the production of disease in an offspring by
a parent who is afflicted by the Münchhausen syndrome. The child had
repeatedly been given doses of the drug promethazine sufficient to induce
seizures and loss of consciousness; the mother herself had obtained medical

attention for *factitious* strokes and seizures in the past. Why Polle? The eponym comes from the given name of the real Baron Münchhausen's only child, who died of unknown causes in infancy. Cases of Polle's syndrome have also been reported yearly since its original description.

Meanwhile, back at Bellevue, I had confessed to the house staff that I had some doubts as to details of Mr. Malone's history but did not venture the alternative diagnosis until we had examined the patient together. We trooped upstairs, were filtered through the three sets of steel bars, brushed shoulders with huge cops and fierce Rastafarians, and arrived at Mr. Malone's bedside.

The story of his illness which Mr. Malone told to the group that morning was in every point identical to that told Mike the night before. Since we seemed appropriately impressed by his social and academic achievements, his basic truculence dissolved somewhat; he began to boast of his computer skills. Directing the questions back to the onset of his disease, and contrasting his athletic youth with his present illness, we soon became convinced that while he may at one point or other have visited MIT he had never been a student there. He was under the mistaken impression that the "rowing team" wore crimson, that the school's best-known economist was named Galbraith, and that Salvador Luria was the Institute's motto.

We discussed Bartter's syndrome, the specifics of which he knew somewhat better than those of us who had not recently done some quick library work. I expressed my admiration to him for having met the discoverer of his disease and innocently asked him what Dr. Bartter looked like. It was no overwhelming surprise to me when he described Dr. Bartter of Dallas (*sic*) as a tall, gaunt man with black hair, a description that differed in every particular from my recollection of that fine physician, with whom I shared the podium at several clinical meetings before his death in 1983.

Physical examination quickly revealed that this large specimen was as healthy as an ox. By means of techniques designed to spot malingerers in the U. S. Army, it was not too difficult to demonstrate that his muscular weakness was feigned. In asides to the house staff, I had pointed out the erroneous "pearl" that one-sided deafness was a feature of Bartter's syndrome. This point is easily established by putting a stethoscope to the *patient's* ear, clamping one of the two tubes with a hemostat, and twisting the scope like a telephone wire. It takes a fast malingerer indeed to reckon the geometry of sound and to report with consistency which ear fails to report the sound of taps at the business end of the stethoscope. Mr. Malone tried to oblige us by offering us the new symptom of deafness, but he was no better in his estimates than were the poor recruits of Fort Dix who wanted to dodge hot days on the firing range.

We left the bedside, having assured Mr. Malone that he would obtain appropriate treatment. We ordered him placed on a regular diet with no potassium supplements, and suggested that he be observed for unusual behavior. Since prisoners cannot readily take unprescribed laxatives or diuretics, it seemed to us that if his potassium were maintained at stable levels for a week, we could readily eliminate the possibility of Bartter's syndrome. Meanwhile, one could call the various doctors he had consulted in the Bay Area to see if his potassium had *ever* been low.

The time had come to write the diagnosis on the chart. The hunch seemed right: Münchhausen syndrome it was.

In 1785, there appeared an anonymous book in London entitled *Baron Münchhausen's Narrative of his Marvellous Travels & Campaigns in Russia*. The publication was only eighty pages long and sold for a shilling. It told tall tales of the baron's exploits with sword, musket, and horse. The Age of Reason stopped at the borders of Russia and Turkey: Beyond the eastern frontier lay a never-never land where time and space were flexible enough for the baron to describe the inhabitants of the moon. The book immediately became popular, and in its fifth edition was translated into German by G. A. Burger; this is the version that became an international classic. It was a nineteenth-century biographer of Burger who finally revealed that the author of the by-now-famous book was a certain Professor Rudolph Erich Raspe (1737–1794). Raspe would have been a suitable roommate for Mr. Malone.

Raspe was a German geologist and geographer who was educated in Göttingen and Leipzig, centers of the German Enlightenment. In his youth he met the voluble Karl Friedrich Hieronymus, Freiherr von Münchhausen (1720–1797), a veteran of campaigns in Russia and Turkey. The spark of adventure was lit at the tableside of the Freiherr. By the age of thirty, Raspe had gained a full professorship at the University of Cassel and become its librarian. As adroit in the salon as at the earth sciences, he became keeper of gems for the Landgrave of Hesse. This job, which involved the trading of jewels, curios, and artifacts on behalf of the princeling, permitted Raspe to travel about Europe as a kind of roving antiques dealer. Raspe did not conduct his various transactions entirely for the benefit of his patron and pocketed the proceeds of sales from the Landgrave's collection of gems. His embezzlement discovered, Raspe escaped to England, where he became involved in various industrial and mining enterprises. By and large he made his living by his wits and guile. While down and out in London he resurrected the table talk of his youth in the form of the baron's tall tales of the east. But real-life shenanigans came first. He obtained employment with a Sir John Sinclair, upon whose holdings Raspe claimed to have discovered veins of precious metals. When it turned out that Raspe had seeded the mines himself, he was again forced to flee westward. He absconded to Ireland, where he spent the remainder of his life in literary pursuits.

Raspe immersed himself in works of legend and the imagination, and from his Irish exile made his second contribution to our cultural history. It was Raspe who first called to continental attention certain writings attributed to Ossian, the ancient Celtic bard, son of Finn mac Cumhail, the warrior hero. They had recently (1760) been "rediscovered" by James Macpherson and denounced, correctly, as pure fakes by none other than Dr. Samuel Johnson. Fake or not, thanks to Raspe and other devotees, Ossianic myths of Death and Transfiguration fed the fantasies of the early Romantics in England, France, and Germany. Ossian was one of Napoleon's heroes and to these proto-Wagnerian legends can be traced much of what is wild and aggressive in European art of the nineteenth century. Raspe, Macpherson, Ossian, and Münchhausen—names that stand for lie and legend—can be carved on the tombstone of the Age of Reason.

If the tales of Münchhausen are on the flip side of the Enlightenment, the case histories of the baron's syndrome can be found as a sorry footnote to modern medicine. It could be argued that what Mr. Malone and his ilk exploit—for whatever sad personal reason—is our *scientific* method of diagnosis. Many of the recent cases of Münchhausen syndrome have been nurses, technicians, scientists, and engineers who know enough to read the popular medical columns or the *Merck Manual*. They fabricate illnesses, like Bartter's syndrome or diabetes, that can be "documented" by analyses of serum or urine. They are the performance artists of laboratory medicine, suffering on their own bodies, or by proxy, the pain of creating an abnormal blip on a hospital machine.

As expected, Mr. Malone did perfectly well in the absence of therapy; his potassium and aldosterone levels were entirely normal. Calls to the Bay Area confirmed that he had been seen in most of the university clinics. His muscle weakness and low potassiums had been investigated many times for possible Bartter's syndrome but he had always failed to return for definitive diagnosis. The California doctors did not know then, nor do we at Bellevue know now, whether Mr. Malone took diuretics, laxatives, or purges to change his potassium levels. Nor is it really our business. Mr. Malone came as a patient, and, unhappily, we know of no cure for this sort of disease; we can only prevent ourselves from hurting him more than he has hurt himself. I am sure the performance will be repeated, as I told the house staff when he was sent back to the criminal justice system. I also read to them the words of Roy M. Meadow, who dedicated his paper on "Münchhausen Syndrome by Proxy" to "the many caring and conscientious doctors who tried to help these families and who—although deceived—will rightly continue to believe what most parents say about their children, most of the time." I will continue to believe what most patients say about their diseases, most of the time.

19

THE DOCTOR WITH
TWO HEADS

The Seine flows west through Paris past monuments to ineffable glory and reminders of unspeakable deeds. The river arches for about five miles through the center of the city and its banks yield views that define the official text of French history. Parisian children are taught that the towers of Notre-Dame sing the Age of Faith, while the Louvre and the Tuileries proclaim the Age of Kings. The grand dome of the Panthéon is a eulogy for the Age of Reason, and behind the Invalides flutter the flags of Napoleon's lost empire. The Eiffel Tower is an exclamation point of industrial mettle, and nowadays the victories of Foch and de Gaulle are bespoken by the Arc de Triomphe.

Strolling on cobblestones beneath weeping willows, under the arches of an ancient bridge, among fishermen, wandering lovers, and widows walking dogs, one looks up over the traffic-laden quays to see those glorious chunks of architecture against a mackerel sky. In this country of Descartes, spire and obelisk mark the distant ordinates of faith and reason. Especially after a simple five-course lunch and a liter or two of agreeable wine, the prospect of this most beautiful and horizontal of cities is enough to explain why Germans use the expression *"Glücklich wie Gott in Frank-reich."* Who would not be happy as God in France? Here, from the silver river's edge, all seems *luxe, calme, et volupté.*

But from time to time the river rises. A false spring in mid-February brings weeks of warm drenching rain; the snows melt early in the east. The water turns an angry brown, and the current becomes so strong that violent froth churns at the foot of bridges. The fishermen, lovers, and widows are displaced to the stone quays which rise a full story above the normal waterside. Benches, railings, and plantings at the water's edge disappear in the flood as the Seine rises by a yard or two. The willows look like soaked spaniels. The tourist boats, the *bateaux-mouches*, can no longer slide under the arches of the Pont Marie and the expressways disappear into brown soup. Uprooted trees and sidings float downstream; errant gulls hitch rides on this flotsam to the western sea. The muddy waters strand houseboats and barges almost in midstream; riverside commerce comes to a halt. One might say that the Seine also rises. We know that there will be no flood here in the heart of the city—there has been none in this century—but can we be entirely sure?

When the river rises, the stroller or jogger is also diverted to the upper quays. Safe from the rushing waters, he finds less congenial running room: the traffic is aggressive, the pavement is fouled by dogs, a hostile concierge hoses down his shins. The dead winter sky hangs low. But the slower pace has its compensations; one has time to notice monuments to the past less imposing than the Eiffel Tower, sermons in stone less sublime than Notre-Dame. This nation of archivists and concierges has kept its ledgers. Stone tablets on quayside façades date the comings and goings of the great, their habitations are labeled as carefully as roses in the didactic Jardin des Plantes. We are informed where on the quai Voltaire lived a troubled Oscar Wilde, and where Baudelaire played with hash. Under a skylight on the quai d'Anjou, Daumier inked his stones while Courbet held court nearby. George Sand entertained Chopin on the first floor of a Left Bank hotel; down the road Borges paid a call. Science is honored as well: the statue of Lamarck (1744–1829), "discoverer of the theory of evolution" sits brooding over the untidy Seine; farther up an *allée* we find Buffon (1707–88), "father of natural history," who seems to have turned his back to cope with pigeons.

The ambiguous history of Parisians between 1939 and 1945 has not escaped the mason's notice. Tucked into a cryptlike embankment at the rear of mighty Notre-Dame is the Memorial to the Deportation; the sepulcher reads: "To the two hundred thousand French martyrs who died in the camps of deportation, 1940–1945." When the brown Seine rises, the memorial—but not the cathedral—becomes inaccessible. Notre-Dame faces the imposing Prefecture of Police and its adjacent prison of the Terror, the Conciergerie. Under German occupation the French flics by and large did what they were asked by their Nazi counterparts, a sort of professional courtesy among concierges. The *allemands* had help in rounding up 200,000 Frenchmen. The clandestine newspaper *Combat* reported that the German authorities "congratulate the French police, who in collaboration with the

German police permitted the arrest of the guilty [a *permis l'arrestation des coupables*]."

Who were the guilty? On the walls of a Jewish welfare hostel, the Fondation Fernard Halphen on the Île St.-Louis, a tablet reads: "In memory of 112 inhabitants of this building, among them 40 children, deported and killed in German concentration camps. 1940." Beneath a similar memento to infanticide in the Marais has been added the admonition: "N'OUBLIEZ PAS!"

The French have also not forgotten their nobler half. Small marble plaques with the *tricolore* mark sites at which French Resistance fighters died in the course of the street fighting that freed Paris in advance of the Allied armies in August 1944. On the wall of the Prefecture—among the first strong points to revert to the Resistance—is affixed a marble and gold tablet:

> HERE WAS RECEIVED
> ON AUGUST 24 1944
> FROM A LIGHT AIRPLANE
> OF THE SECOND ARMORED DIVISION
> THE MESSAGE OF GENERAL LECLERC
> TO THE PARISIAN RESISTANCE
> "HOLD ON, WE'RE COMING"
> ["TENEZ BON NOUS ARRIVONS"]
> DROPPED BY CAPTAIN JEAN CALLET
> AND LIEUTENANT ÉTIENNE MANTOUX.
> KILLED ON THE FIELD OF HONOR

The battle did not really end in 1945. Fratricide came to an end only after Pierre Mendès-France and Charles de Gaulle pulled out of Indochina and North Africa: the last scar of self-laceration is etched on the wall of No. 25, quai des Grands Augustins: "Here lived Dr. GEORGE FULLY, member of the Resistance, deported to Dachau, man of Liberty and Justice. Assassinated June 20, 1973."

It may seem a little naive to read these quayside inscriptions as a guide to the cultural geography of two Frances: the republic of Liberty and Justice and the fiefdom of the concierge. But we have learned from the new doyens of French social thought—Foucault, Barthes, Derrida—that one can decipher the signs of a culture from the spaces assigned to words and to words that are never spoken. In this sense the apparently random inscriptions above the surging Seine might be read as a promise that *nothing* will be forgotten. Come hell or high water, Captain Callet, Dr. Fully, and the forty children from the Île St.-Louis are part of the discourse that France conducts between enlightenment and the cops.

Dr. Fully is not the only doctor remembered on the quays. There is a whole museum devoted to them on the quai de la Tournelle, close to the ritzy restaurant La Tour d'Argent. When my wife told me about the Musée de l'Assistance Publique I was pretty skeptical about the enterprise. After all, the Assistance Publique is simply the Parisian version of the Health and Hospitals Corporation of New York City: both administer the public hospitals and clinics of their city. In fact, Bellevue Hospital—where I work—and its sister institutions serve functions based on older French models, the Hôpital Dieu and the Salpêtrière, which care for the sick and the mad. I wondered what could possibly be displayed in this museum of social welfare other than the usual assortment of old microscopes, ancient ambulances, and stuffy portraits of forgotten professors.

With the quays awash, the Musée was on my new running path, and I found the familiar logo of the Assistance Publique on a blue banner that hung from a splendid seventeenth-century mansion, the Hôtel de Miramion. Needless to say, the doors were closed at eight-thirty in the morning. But mounted in an old display box to the right of the entrance was a poster that knocked my eyes out.

Brushed in acid colors, with a touch of bravura that evokes the Montmartre of Toulouse-Lautrec, the painting shows a fellow in top hat and white gown about to buzz the creamy *poitrine* of a helpless woman with an electrical gadget that could have been assembled by a junior at Bronx High School of Science. A fast reading of this sexually charged tableau might yield the message of womankind at the mercy of man and his infernal machines. Who painted this scene, man or woman? The poster was weatherworn and I could hardly decipher the signature: CHICOTOT. Was the fellow with the top hat a doctor? If he *was* a doctor, why was he top-hatted indoors? What was a stiff devil like that doing to merit a share of quayside immortality with Dr. Fully, man of Liberty and Justice?

It took several visits, some hours in the library, and the help of Nadine Simon, curator of the Musée de l'Assistance Publique, to work it all out. The painting, signed "CHICOTOT, Georges 1907," is a self-portrait in which the doctor-painter shows himself embarked on "the first trial of X-ray therapy for cancer of the breast." It is also a smashing painting, in which the pigments splash unnatural green, red, and yellow highlights over the

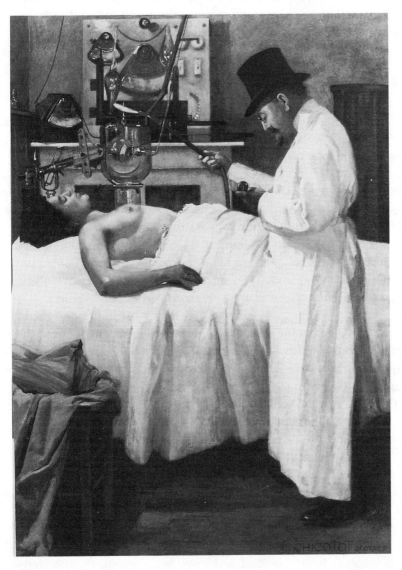

G. Chicotot, *The First Trial of X-ray Therapy for Cancer of the Breast*. Musée de l'Assistance Publique, Paris.

intricate apparatus: a postimpressionist view of new science. The canvas also fulfills the prediction of a contemporary critic, who wrote in *Le Correspondant médical*, "One sees how precious this document will be in years to come, when a writer of a future generation will trace the history of this novel form of therapy." Well, to this "écrivain d'une génération future" the canvas is something more; it is not only an icon of clinical research but also an emblem of the liberal, bourgeois republic of Clemenceau, the republic of doctors.

The picture can be read on several levels, but the scene is unequivocal. Dr. Georges Chicotot, head of radiotherapy at the Hôpital Broca, shows himself treating cancer with X-rays. In his left hand he holds a watch to time the exposure; in his right he holds a sort of extended Bunsen burner that spouts flame from its tip. He is heating the vessel that holds the generator, the Crookes tube. The X-rays are focused on the patient's breast by a glass cylinder. The ominous electrical apparatus on the mantelpiece is simply a transformer, and the two vacuum valves on either side of the tube regulate the current. The Crookes tube and its enclosing vessel are painted in eerie green, yellow, and orange. The woman, who is either sedated or oblivious, is undressed. Her corset and dress are shown on a stool at the left. She is no *jeune fille*; a wedding ring is shown on her right ring finger. All of the composition lines of the complex painting lead to her right breast; she is a beautiful woman with a fatal disease. Chicotot does not look at her—his eyes are on the watch—but the tube in his right hand is aflame. He wears a top hat and huge apron, perhaps because—to quote a description of this work from the *Presse médicale* of 1932—"all doctors of the time were recognizable in their laboratory by their top hat [*chapeau haut de forme*] and white apron."

This picture must be the only one in the history of experimental medicine in which the doctor produced not only data but art! Chicotot was what the French call *bicéphale*; he was a doctor with two heads. In the land where the form of a hat follows the function of the wearer, he was entitled to two. An honored graduate of the École des Beaux-Arts, he won several medals for historical paintings which he exhibited at the annual Salons. His paintings were highly finished and based on an extraordinary interest in anatomy, which he had taught as a prosector in the School of Practical Anatomy. In this unique Parisian institution, young painters and doctors alike learned the disposition of muscles, tendons, and fasciae. One métier led Chicotot to the other, and he entered the École du Médicine, from which he earned his medical degree in 1899. He obtained an externship—a rare prize—and soon was launched on a career in the early days of radiology. His work on X-rays at the Hôpital Trousseau gained him another medal, this time from the Academy of Medicine. Meanwhile, his paintings at the annual Salons turned more and more to medical themes, provoking *Le Correspondant médical* to claim that his gripping scenes "attracted the attention of the general public, which is fascinated by the subject of our art." They also constitute a pictorial autobiography.

A self-portrait of 1900 shows Chicotot soldering a homemade vacuum tube; he is in shirtsleeves and wears a white apron, but no top hat. In the Musée de l'Assistance Publique is another splendid Chicotot canvas of 1904: it shows a Dr. Josias inserting an airway into the throat of a small child with diphtheria. The infant is perched on its mother's lap and its head is supported by the professor's assistant. Eight other young doctors, among them Chicotot, watch the delicate maneuver. They are in street clothes, protected by white aprons: not a top hat is in sight. In 1905, Chicotot exhibited an evocative painting entitled *Autopsy: At the Dawn of the Twentieth Century*. Dr. Tollemer, a pathologist, is shown aspirating the thoracic cavity of an infant in order to establish a bacteriologic diagnosis. Chicotot has painted himself in the act of plating out the cultures. A nurse, coiffed and capped, surveys the scene. Both doctors are in white aprons; neither wears a top hat. Indeed, not one of the docs in Chicotot's paintings before 1907 wears a top hat! Nor for that matter do any of those cool, reportorial canvases sport a flash of unnatural color or hint of avant-garde brushwork. Before 1907, Chicotot was the careful observer, the *aspirant* of the Beaux-Arts, the prosector of practical anatomy. With his X-ray picture, dominated by an eerie apparatus brushed in acid colors, he moved into our anxious age of gadgets and Freud.

Top hat and bare breasts suggest the brothel, not the frontiers of medicine. How common are these images? How often did the French, whose talents with respect to the depiction of bare female flesh cannot be said to lag behind other nations, show chaps in top hats and street clothes having traffic with nudes? To answer these questions I conducted a fast field survey in the new Musée d'Orsay, which, a few laps down the quay from the *musée* devoted to the Assistance Publique, has been built in the shell of a spectacular railway station of the Belle Époque. Its fussy renovation has permitted France to empty the basements of the Louvre and to display its nineteenth-century holdings in unquestioned grandeur. I made a checklist at the Orsay of some 286 paintings, dating between 1840 and 1903. While I may have missed some—we're not talking pastels or sculpture—I was able to find 89 canvases that featured bare-breasted ladies; the search was no great hardship. Among all those paintings, those nymphs, muses, and goddesses, only three showed men in contemporary clothing in the presence of nudes.

The first is the well-known *Artist's Studio* of Gustave Courbet (1855), an allegory in which the painter shows himself in his studio surrounded by persons—real or symbolic—important to his career. In the picture, Courbet sits in a chair from which he daubs at a canvas with a long, flickering brush. He is at work on a large landscape and his back is turned on a solid, shapely nude—his muse of truth—who regards his work with adulation. Her *déshabillé* may have been the inspiration for the discarded street clothes of Chicotot's patient. Courbet sports a painter's smock: no top hat here. The rear of the studio is filled with a crowd of his friends from the *vie bohème* and the politics of the time: Baudelaire, Champfleury, Proudhon. It was Proudhon who wrote the crabby pamphlet "The Pornocracy of Women in Modern Times" (1875) and who was persuaded that women had only two functions, that of housekeeper and that of prostitute. Proudhon's attitudes remind me of statistics presented by the cultural historian Theodore Zeldin that in 1882 there were seven practicing women doctors in France, and in 1903 ninety-five; by way of contrast, there were 15,000 prostitutes registered in Paris alone, and between 1871 and 1903 there were 155,000 on the books; 725,000 others were arrested for practicing without a license, so to speak.

Proudhon rated the intellectual and moral value of women as one-third that of man. This paragon of French socialism formulated the principle that is so well illustrated by Courbet in the painting: "Man is primarily a force for action, woman for fascination." His pronouncements resemble those Mort Sahl parodied a century later: "A woman's place is in the oven."

The next painting to show a nude in the presence of clothed men is Manet's *Déjeuner sur l'herbe* (1863). This picnic scene might seem to be off the track of Courbet and Chicotot, but we can look at it as another example of the artist-and-model genre worked by Courbet and Chicotot. Manet shows us his model in the outdoor studio of the forest of Fontainebleau: also seated on discarded underclothing, she stares at us from the canvas. The two men neither look at nor speak to her. The artist, whose métier we know from the soft, bohemian hat he wears, grips a long cane in his hand.

G. Courbet, *The Artist's Studio*. Musée d'Orsay, Paris. © Photo RMN.

E. Manet, *Déjeuner sur l'herbe*. Musée d'Orsay, Paris. © Photo RMN.

The third of these paintings is more to the point. The canvas is hard to find in the quirky museum, where it is jammed into a corner of the space devoted to the arts of the Third Republic. The picture can by no means be said to make it to the finals against Courbet or Manet, and it would certainly lose in the quarterfinals to Chicotot. Painted for the Salon in 1887 by Henri Gervex, it is entitled *Avant l'opération* and shows Dr. Péan of the Hôpital St.-Louis at an operating table. A well-painted young woman, anesthetized and undraped from the waist up, lies before the doctors, her ginger hair spilling over the coverlets. Péan is demonstrating before five spectators the use of a new hemostat. He holds the pointed instrument over the sleeping woman. Other surgical paraphernalia lie at the head of the table. The surgeon and his assistants are in street clothes. In 1887, to quote the *Presse médicale,* one "wore the same outfit in the operating room, at the bedside, and at the autopsy table!" No top hats here, either, but a nurse is in attendance.

In fact, a second, clothed female is depicted in each of the three paintings at the Orsay. In none is the artist or doctor alone with his unclothed subject: male and female chaperones tell us that a professional interpretation is appropriate. Artist and doctor alike had access to the unclad female body, both having earned that right by long study in the prosectorium where, as in this poem by Professor J. L. Faure,

> *C'est que son but est noble, et ces débris immondes*
> *Et ces lambeaux sanglants sont les pages fécondes*
> *Où l'on apprend à lire au grand livre du sort*
> *Et l'on connait la vie en fouillant dans la mort.*

> [They pursue noble aims amid foul remnants
> and bloody dressings which serve as fertile texts
> from which one learns to read the great book of fate
> and to understand life from a rummage with death.]

The iconography of Chicotot becomes a little clearer if we remember his standing as a *bicéphale*. We can assume that he shows himself wearing a top hat as a badge of entitlement, a sign of higher office. The hat, serving as symbolic chaperone, tells us that the doctor's tinkering is sanctioned by learning. Chicotot also wears his topper as a sign of class. Following the lead of aristocratic dandies, the bourgeoisie sprouted the *chapeau haut de forme* on all the playgrounds of the class game. At the Musée d'Orsay, I found scores of paintings in which Frenchmen are shown in toppers at the races, at balls, at assemblies, on the boulevards. If crown and scepter were the symbols of power in the Age of Kings, so top hat and cane became the objects of rank in what Roger Magraw has called the "Bourgeois Century" in France (from 1815 to 1914).

One can see in another painting at the Orsay that the successful artist as well as the successful doctor was entitled to his topper. The canvas, again by Henri Gervex, is entitled *A Session of the Jury of Paintings* and shows Gervex with other established figures of the Beaux-Arts selecting the paintings for the Salon of 1886. These doyens of fine arts are clustered around a large canvas on an easel. Most of the dozen or so of these experts—all in top hats, worn indoors—have raised their canes in approval; they are giving the official "thumbs-up" sign in the arena of Salon painting. Gervex, the *chef des artistes*, seems to be telling us that when painters have arrived they put behind them the things of childhood. Top-hatted, they raise canes.

Early in the nineteenth century, "anyone who could afford to pay for his own funeral" was considered a bourgeois, but as industry spread and the tribe increased, other descriptions applied. The French sociologist Édmond Golot summarized a major requirement: "A bourgeois had to be able to perform his job in bourgeois costume, so that manual or dirty physical work was unacceptable." Only painters and doctors were permitted to be seen in the alternative costume of smock or white apron.

The historian Jerrold Seigel reminds us in *Bohemian Paris* that when Rodolphe laments over his dead Mimi in Henri Murger's *Scènes de la vie de Bohème*, he cries, "*O ma jeunesse! C'est vous qu'on enterré*" ["O my youth! It is you that I bury"]. We might say that Chicotot's canvas is an active lament for his youth as he realizes he is no longer the young *aspirant* but now a *chef du service*. His top hat signals good-bye to his bohemian past. Doctors and painters alike were expected to have passed through a bohemian phase of training, a youthful fling at eccentric dress. Murger, who in 1849 introduced the legend of bohemia to our culture, defined its limits as "bordered on the North by hope, work and gaiety, on the South by necessity and courage, on the West and East by slander and the hospital." Atelier and hospital were the stage sets of bohemian theater.

The yearly hospital balls were as frisky, bawdy, and reckless as those held by the artists. But in their mature years, doctor and artist alike willingly assumed the symbols of bourgeois achievement: top hat and cane. In his X-ray picture, Chicotot shows himself not as a bohemian, but as the mature medical scientist, admitted as by seigneurial right to the intimacy of an

H. Gervex, *A Session of the Jury of Paintings*. Musée d'Orsay, Paris. © Photo RMN.

uncthaperoned woman. He does not require a chaperone because his rank and profession (hat and paraphernalia) place him *hors de passion*. He has passed the rites of study and apprenticeship and with Bunsen burner in hand can make the woman better. Perhaps he can bring this Mimi back to life! This rescue fantasy of oncology without mutilation makes an uncomfortable picture, charged as it is with Freudian images and touched by death. One also senses in this farewell to bohemia a painterly homage to a new, freer art, ablaze with the colors of the fauves and the broader highlights of postimpressionism.

Alfred Delvaux explained the need of the mature bourgeois for his alter ego, the younger bohemian: "Because he started out by being you before becoming himself, because he had a heart before he acquired his tummy, because he had debts before he had bonds, because he had long hair before he had a trimmed lawn, because he had a mistress before he had a wife. He is a conclusion to a book of which you [the bohemian] are the preface." This sympathetic view of the dialectic between bourgeois and bohemian might go far to explain why accountants from Teaneck and underwriters from Greenwich rise from their expensive seats in standing applause as Victor Hugo's revolutionaries wave the *tricolore* over the Broadway version of *Les Misérables*.

Many paintings, the *Arnolfini Marriage* by Jan van Eyck for example, are signs of a contract between the persons depicted. We could say that Chicotot's painting constitutes not only a contract between doctor and patient, but also an uneasy compact between the two Chicotots: between the young bohemian and the older bourgeois, between the artist and the scientist. As far as I have been able to establish, this was his last painting of note: it marked the end of his career as a *bicéphale*. The chief of radiology had displaced the painter. It is not a happy painting; the pleasures of Courbet or Manet cannot be found in this work. X-ray therapy is not foolproof as a cancer cure: oncology is not a carefree enterprise. But as Manet's good friend Émile Zola wrote in 1892: "Does science promise happiness? I do not believe so. It promises truth, and it is questionable whether one can ever be happy with the truth."

Some truths *are* happy, however. My father's old textbook from which he learned surgery, published in 1911, points out that by and large most women with breast cancer will die within two years after diagnosis. *CA*, the American Cancer Society journal, of January 1988 gives an overall five-year survival rate of more than 70 percent. Chicotot in his *chapeau haut de forme* and the other scientists of the bourgeois century played no small role in this reversal of fortune.

Zola and the democrats of the Third Republic were major champions of the new scientific medicine. In the year that Chicotot painted his X-ray picture, Georges Clemenceau was premier of France. Clemenceau and his radical republicans had come to power thanks in no small measure to their role in the vindication of Alfred Dreyfus. The Dreyfusards, among them Zola, Drs. Naquet and Colin, and most experimental scientists, grouped themselves around the Ligue des Droits de l'Homme. Their secular, antimilitary sentiments were based on the firm conviction that social and scientific progress were linked. Zola, who had used the experimental method outlined by Claude Bernard as his model for "The Experimental Novel," wrote a medical novel entitled *Le Docteur Pascal*. The hero is patterned on many of the clinical investigators whose names we now know by their eponyms: Gilles de la Tourette, Babinski, Charcot. Zola summed up the scientific faith of the era:

> I believe that the future of mankind lies in the progress of reason through science. I believe that the pursuit of truth through science is that divine ideal to which man should aspire. I believe that all is illusion and vanity which is not among the treasures of verifiable truth that must be slowly acquired. I believe that the sum of those truths which are bound to increase continually will eventually give man incalculable power and equanimity, if not happiness.

Clemenceau was a doctor, and had begun his political career as mayor of Montmartre; his convictions were forged in the crowded clinics of the very poor. Unlike the professors of the hospitals, the doctors on the front lines of Paris medicine in the clinics of the Assistance Publique were on the lowest rungs of the top-hat ladder. Zeldin quotes a Dr. V. Macrobius (*sic*) who in 1889

> protested against the ridiculous fee of one franc per annum paid for working in the medical service for the indigent, objecting to doctors being pushed to the very bottom of the social ladder and ironically declaiming: "We do not dispute the first rank to the magistrature, clergy or army. Nor do we wish to place ourselves on the same level as engineers, actors, painters, architects or sculptors. We ask only that we may occupy a middle rank, more or less: we would like, for example, to be classed between the solicitor and the photographer."

In or out of power, the party of Clemenceau had the support of many such doctors: ten of the thirty-four deputies with whom the radical republicans began in 1876 were physicians. Together with other anticlerical, antimilitary liberals they started treating illnesses of the body politic. They successfully mandated universal education, introduced divorce, separated church and state, and extended suffrage beyond the propertied classes. They also fought the battle of Dreyfus against the combined forces of the Church, the army, and an intransigent right.

The bourgeois, meliorist reformers of the Third Republic were attacked by populists of the right and of the left. National socialism of the anti-Marxist kind was a French invention, and Proudhon was its prophet. The line extends through Louis-Ferdinand Céline to the flics of the occupation. Aristide Bruant, singer, man of politics, and *fin de siècle* rabblerouser, attacked the bourgeoisie from the left, running for office with the promise to fight "all the enemies of capitalistic feudalism and cosmopolitan Jewry." From the right, Drumont complained in his *La Libre Parole* (June 25, 1895) that "besotted by the prostitute, robbed by the Jews, menaced by the workers, the Voltairean and masonic bourgeois begins to perceive that he is in a bad way. . . . And all the corruptions he has sown are rising up like the avenging furies to push him into the deep."

The brown waters of the raging Seine have always been there to overflow the borders of reason. On my desk is a copy of *Le Petit Journal* of February 27, 1898. Its red-white-and-blue cover shows a "heroic" Major Henry—the Oliver North of the Dreyfus Affair—confronting Lieutenant Colonel Picquart, the archivist whose evidence was to vindicate Dreyfus. Major Henry flings down the gauntlet: "You have lied!" he tells Picquart. The text describes Henry as "brave and loyal, a child of *real people*, a simple soldier; and now despite courage, energy and devotion to country forced to respond to the likes of Lieutenant Colonel Picquart." When the Dreyfus verdict was reversed, Clemenceau and the republic of doctors made Picquart their minister of war.

Dr. Chicotot and his top hat, Dr. Clemenceau and Lieutenant Colonel Picquart remind one that *bourgeois*—noun or adjective—need not be pejorative. Jean Rostand quipped that scientific research is the only form of poetry that is supported by the state. The poetry of French clinical research was supported by the bourgeois republic and opposed by its enemies on the right and left. Unfortunately, the republic of doctors with two heads, the meliorist brotherhood of the bourgeoisie, was laid low by the epidemic of nationalism that broke out in 1914. Clemenceau was no exception, and

Chicotot served voluntarily as a combat officer, although exempt from service for reasons of age and profession. There have been no *bicéphales* of Chicotot's measure in France ever since.

Running toward the statue of Lamarck, nowadays, one trots along the allée Claude Bernard, which is lined by early-blooming forsythia. Crossing the wide boulevard in back of the Jardin des Plantes, one arrives at the great, domed Salpêtrière, a world center of neural science and healing. In the fourth courtyard of this monument to sound reason and fit architecture is the Charcot Library. At its entrance is a huge, realistic tableau painted by Tony Robert-Fleury in 1876. It shows Dr. Pinel in revolutionary hat and cane presiding over the unchaining of the mad. The central figure is a beautiful, partially clad young woman with the detached look of the very lost. A few other women lie about, some in Salon poses of the chronically distraught. Wardens are in the process of breaking the chains: gratitude and justice are the message of the canvas. The young woman in the center of the painting is posed in a gesture that reappears at the Musée d'Orsay, where a slick marble statue by Ernest Barrias (1895) shows another beautiful young woman lifting her veil to display her *poitrine*. The statue is entitled *Nature Unveiling Herself to Science*.

In the course of the bourgeois century men of Liberty and Justice were in no doubt that when Nature was revealed to Science she would turn out to be beautiful and just, that when the chains of the Conciergerie were struck, the tablets over the hospitals of the land would read LIBERTÉ, ÉGALITÉ, FRATERNITÉ.

Emerging from the Salpêtrière, one comes to the quays again, where a school bus discharges a score of school children off to frolic in the Jardin des Plantes. Oblivious to the river, which poses no threat, their happy faces bespeak the new Paris of many races; their parents have come from North Africa, from Haiti, from Vietnam and the Ukraine, as well as from the Île-de-France. In fluent French that any Anglophone would envy they chirp the sweet noises of fraternity. They are the end to which the better France has always been devoted. They are heirs to the republic that signs itself—as in the March 1944 issue of the Resistance journal *Combat médical*—"We, doctors of the National Movement against racism and those who have never before been committed to the battle. All victims of the Germans are victims of racism: young Frenchmen threatened by deportation or persecuted Jews will find in us an ally. That is why we join forces with all those of the Resistance to hasten the day of victory when our land will be free."

T. Robert-Fleury, *Dr. Pinel Unchaining the Mad*. Bibliothèque Charcot, Collection Hospital Salpêtrière, Paris.

L. E. Barrias, *Nature Unveiling Herself to Science*. Musée d'Orsay, Paris. © Photo RMN.

20

WORDSWORTH AT THE
BARBICAN

The Barbican Arts and Conference Centre, a glum assembly of concrete high-rises and bunkered flats, overshadows the City of London. Oldest of London's districts, the City for several centuries presented a harmonious skyline in which no structure competed for attention with the splendid dome of St. Paul's Cathedral, designed by Christopher Wren. Pierced only by the occasional steeple of one of Wren's lesser masterpieces or by a Regency cupola, the townscape remained open to the English sky and its infrequent gift of sunshine. Seen at dawn from the distance of Westminster Bridge, the City prompted Wordsworth's claim that

> *Earth has not anything to show more fair:*
> *Dull would be he of soul who could pass by*
> *A sight so touching in its majesty:*
> *This City now doth, like a garment, wear*
> *The beauty of the morning; silent, bare,*
> *Ships, towers, domes, theaters, and temples lie*
> *Open unto the fields, and to the sky;*
> *All bright and glittering in the smokeless air.*

> "Composed upon Westminster Bridge"

That description does not apply today. Although some of the majesty remains, the grime of the Industrial Revolution, the incendiary bombs of the Luftwaffe, and the towering cranes of modern planners have turned great parcels of the City of Wren and Wordsworth into a Houston-on-the-Thames. One might, of course, argue that the modern city isn't all bad news. Whereas Earth may not have had anything to show more fair than the London of 1802, it certainly may be said now to have places to show that are cleaner and healthier: the London of 1990, for one. By my reckoning, even the grim Barbican has at least two desirable features.

First, the planners have encased a rich cluster of cultural treasures within those concrete bunkers. The Arts Centre houses not only the London Symphony Orchestra but also the Royal Shakespeare Company. There are large and small concert halls, theaters, and cinemas, two exhibition salons, art galleries, the Museum of London, a botanical conservatory, and a variety of meeting and practice rooms. This mall of culture is flanked by associated shops, pubs, and restaurants that have been neatly apportioned among the concrete piazzas and dim loggias.

The second feature is probably unplanned. No doubt on the basis of the questionable experiments of Le Corbusier in sunnier climes, modern architectural canon decrees that its megaliths be hoisted on pylons, among which the pedestrian can amble in the shade. The Barbican, no rebel against canon, has pylons aplenty. Removed from cars and commerce by empty plazas and shop-free arcades, the pedestrian can pick his way among the pylons to avoid the rain but not—alas!—the wind of London in autumn. The avid jogger soon appreciates that the dreary acreage of traffic-free Barbican is perhaps the only spot in central London where he can plod for a few miles without becoming soaked or choked.

On a recent Saturday morning, I was taking a run through the Barbican and became lost among all those pillars. Dodging the drafts, I suddenly found myself in a semicircular corridor, carpeted in red and decorated with an excess of brushed chrome. Signs informed me that I was on level six of nine levels of the Barbican, and it struck me how appropriate it was that this place, so like the one described by Dante, was divided into levels or circles rather than old-fashioned floors or stories. The corridor along which I now loped was faced on one side by glass partitions, through which could be seen the administrative offices of the Barbican. They were brilliantly lit and empty, save for one large antechamber, in which a dozen or so overtly miserable people sat waiting before a Cerberus-like secretary.

Curious as to what the action was in this belly of the cultural beast, I stopped to read a small placard on the door: "AUDITIONS THIS SATURDAY 9:30 A.M. IN ROOM B." Through the glass, I could see that the sad young people were musicians—string players in their late teens or early twenties, with damp black cases containing violins, violas, or cellos lying at their feet or across their knees. The women's wet hair was very short, the men's very long, and all fidgeted with it a good deal. Puddles collected at their feet.

They did not speak to each other and assumed the expected demeanor of poor relatives about to be read out of a rich uncle's will. It was nine forty-five, and the honcho or hiring committee was clearly not on board. As the group waited, they lapsed into positions that now resembled those of outpatients at an oral-surgery clinic. They leafed glumly through damp newspapers.

Their obvious discomfort, wet locks, and clearly precarious position aroused sympathy. I recalled Anthony Powell's comment: "Reverting to the University at forty, one was reminded of the unremitting squalor of the undergraduate existence." That squalor is a function not only of means but also of ways. The young—ambitious, feisty, and filled with single-minded delight in mastery of their métier—are forced to jump through so many mazes, to wait in so many anterooms. How few, if any, of these young talents here on the block will be making a living at the fiddle or cello a decade from now! How long a road to walk for the sake of art! And the career itself: How dependent on luck, on critics, on changing musical fashion! As I was in this avuncular vein, Wordsworth sprang again to mind:

I think of thee with many fears
For what may be thy lot in future years.

"To H. C."

But these charitable sentiments were erased by details of the tableau behind the glass. I noticed that the newspapers in which many of the young instrumentalists were engrossed were of the sort that even the BBC calls the "tits-and-bum tabloids": the *Daily Mail*, the *Daily Express*, the dreadful *Star*, and Rupert Murdoch's *Sun*. Difficult as it may be for an American to abandon the notion that we lead the world in trashy journalism, it must be conceded that the Brits have us beat by a country mile. Pages of scandal and acres of milk-fed flesh are served up morning and evening by these pop tabloids to support the proposition that "*The Sun* never sets on the British rear." Hurricanes may have toppled the oaks of London, the stock markets of the world may have dribbled down the drain, missile treaties may have been signed or broken, but daily the tabloids of England display on page 1 the knees of Princess Di and on page 3 the breasts of a working-class model. This mass assault on modesty, taste, and women in general seems to arouse no great protest on the part of young Albion.

The irony at the Barbican was the sight of all those would-be virtuosos of high art—those future Heifetzes, Yo-Yo Mas, Jacqueline Du Prés—digging their noses into *The Star* with its glossy shot of Marvellous Mandy's bare *poitrine*. Mandy is alleged to have "sauce! But then the magnificent 19-year-old model does come from Worcester. And when she's not posing for the cameras, there's nothing Mandy likes more than reading romantic novels and tuning-in to her favorite telly shows." How unlike the view of women from gentle Wordsworth:

> *A Being breathing thoughtful breath,*
> *A Traveller between life and death,*
> *The reason firm, the temperate will,*
> *Endurance, foresight, strength, and skill;*
> *A perfect Woman, nobly planned,*
> *To warn, to comfort, and command;*
> *And yet a Spirit still, and bright*
> *With something of angelic light.*

"She Was a Phantom of Delight"

Those glorious vocables might seem more appropriate than *The Star* as morning reading for our young instrumentalists. They were, after all, about to make the most sublime noises of our civilization: Mozart rondos, Vivaldi concerti, the unaccompanied cello suites of Bach. Above their heads hung great posters announcing the masters: Rostropovich conducting Tchaikovsky, Jessye Norman singing Puccini, Itzhak Perlman playing Mozart. Oblivious to musical piety, the young paid more attention to Marvellous Mandy and her sisters of the tabloids. It struck me that only Wolfgang Amadeus Mozart might have shown the same preference.

With him in mind, the scene at the Barbican suggested other themes. Mozart, that foulmouthed angel, was the very model of an eighteenth-century Freemason. Skeptical, irreverent, without a trace of conventional piety, he anticipated not only the court manners of our John McEnroe but also the pop spirit of *The Star*. No guardian angel of Arts Centres he! How different in temperament from the CEO of English Romanticism:

> *My heart leaps up when I behold*
> *A rainbow in the sky:*
> *So was it when my life began;*
> *So is it now I am a man;*
> *So be it when I shall grow old,*

> *Or let me die!*
> *The Child is father of the Man;*
> *And I could wish my days to be*
> *Bound each to each by natural piety.*

Wordsworth, Coleridge, and Keats not only waged their Romantic revolution on behalf of "natural piety" but also presided over the expulsion of natural science from the temples of art, beginning with Isaac Newton. The English Enlightenment had been fueled by the science of Newton; the young Romantics seem to have decided that Newton himself must disappear with his Age. In verses that require no annotation by a psychohistorian, Wordsworth recalls his adolescent self peering from a college pillow to see by moonlight:

> *The antechapel where the statue stood*
> *Of Newton with his prism and silent face.*
>
> *The Prelude*, Book III

Once Newton the father was toppled from his Cambridge plinth, a young poet might dare to put the rainbow together again. "O statua gentilissima, del Gran Commendatore," with that prism in the statue's hand! This somewhat Oedipal obsession with Newtonian optics on the part of the Romantics is well described by Marjorie Hope Nicolson in her *Newton Demands the Muse*. At a memorable dinner in 1817, Keats and Wordsworth agreed that Newton was the opposition. He had destroyed all the poetry of the rainbow by reducing it to its primary colors. "Wordsworth was in fine cue," and at the end of the evening, Keats and Wordsworth joined in the toast of Charles Lamb to "Newton's health, and confusion to mathematics!" Wordsworth's verse enlarged on the theme:

> *Whatever be the cause, 'tis sure that they who pry and pore*
> *Seem to meet with little gain, seem less happy than before:*
> *One after One they take their turn, nor have I one espied*
> *That doth not slackly go away, as if dissatisfied.*
>
> "Star-Gazers"

Since natural science, the realm of those who "pry and pore," can yield no satisfaction and since it reduces Nature to mathematics, why not dispense with its study once and for all?

It could be argued that in England art and science parted company somewhere between 1805 and 1820, with the Two Cultures going their separate ways ever since. It cannot be coincidental that Wordsworth and company discovered Nature, or at least the rustic landscape of England, at that moment when the Industrial Revolution was about to change the landscape forever. I have a hunch that the poet's fear of Newton followed the peasant's fear of steam. It was not a prescient vision of future Barbicans but news of a railroad through his home turf at Windermere that provoked Wordsworth to cry: "Is there no nook of English ground secure from rash assault?"

Mad William Blake joined the anti-Newtonians. He was sure that the "Epicurean" philosophies of Bacon, Locke, and Newton were responsible for the dark satanic mills that were defacing England's green and pleasant land. With his characteristic precision of thought he managed to conflate all the enemies of the Romantic movement. A few samples offered by Nicolson suffice:

Item: "The End of Epicurean or Newtonian Philosophy . . . is Atheism."

Item: "God forbid that Truth should be confined to Mathematical Demonstration."

Item: "Nature says 'Miracle,' Newton says 'Doubt.'"

Item: "The House of the Intellect is leaping from the cliffs of Memory and Reasoning; it is a barren Rock; it is also called the barren Waste of Locke and Newton."

Item: "Art is the Tree of Life, Science is the Tree of Death."

Now these unkind aphorisms may be attributed to Blake's imperfect grasp of science, but I'm afraid that the prattle continues today. Cynthia Ozick, also a great fan of the Old Testament, has told readers of *The New York Times Book Review* that art deals with the world of men, science the world of God. O Newton! Thou shouldst be living at this hour.

When I emerged from the swank bowels of the Arts Centre, the rain had stopped. Since the western terraces of the Barbican are only a hundred or so yards from the medical school I was visiting, I headed toward Charterhouse Square. St. Bartholomew's Hospital Medical College is situated at its northern side and has together with the square maintained many of the older graces of the London townscape. Brick and limestone, arch and cupola, lawns and trees are disposed in easy lines; the visual grammar is traditional. In consequence, after crossing Aldersgate Street, the busy traffic artery that separates the Barbican from this oasis of amenity, I found myself on cobblestone beneath the falling leaves. It struck me as ironic that modern science, which is thriving, was housed in conventional structures, while classical music, which is not, is played mostly in temples of confused modern design.

On the steps of the residential hall sat a group of medical students. In age, dress, and demeanor they were indistinguishable from the string players across the road. Some were waiting for the results of an exam to be posted on the bulletin board in the pharmacology building. They cannot have been reassured by the warning that accompanied an earlier posting of grades on that board: "Those with Grade D are going to have to improve their performance if they are not going to sink." The dozen or so students sat under wet leaves, hard rock came from a car radio, and several were preoccupied with Marvellous Mandy in *The Star*. A second tableau!

On this Saturday morning in London, on either side of Aldersgate Street, sat the young of the Two Cultures waiting to have their performances judged. Poised between the safety of school and the hazards of a career, they seemed to be united only by the glitter of pop and the cramps of adolescence. Wordsworth might be called the first poet of modern adolescence; indeed, I've always considered the Romantics to be the laureates of freshman passion, the troubadours of testosterone. No one has gotten adolescent angst better than Wordsworth at Cambridge:

> *Examinations, when the man was weighed*
> *As in a balance! of excessive hopes,*
> *Tremblings withal and commendable fears,*
> *Small jealousies, and triumphs good or bad . . .*
> *Wishing to hope without a hope, some fears*
> *About my future worldly maintenance,*
> *And, more than all, a strangeness in the mind.*

> The Prelude, Book III

Wordsworth's turmoil at Trinity, his fear for his worldly maintenance, remind one that whatever else the Romantic rebellion accomplished, it was coincident with the gradual ascent of the middle classes. Only an aristocracy can afford a cult of the amateur; those who would busy themselves looking about for worldly maintenance, who would wish to rise from mine and mill and field, had better be very good at doing one thing. The middle class was asked to choose art or science, and the Romantics knew which side was theirs. By the 1840s the furrow between art and science had deepened: Wordsworth, Keats, Coleridge, Blake, and Lamb had succeeded in erasing the language of science from the blackboards of culture. And on the other side of the gap, the Royal Society began stripping its rolls of literati and amateurs. From 1847, the pattern was set that guaranteed that for over a century no British youth has been taught *both* art and science at the university level.

It may be a bit facile to trace the divorce in England between the Two Cultures to the Romantic revolution. But the unity of all cultural effort had been the unwritten rule of the Western world from Aristotle to Maimonides, from Avicenna to Spinoza, from the Florentine Accademia to the philosophes. And at the tables of the English Enlightenment, Wren supped with Boyle; Hooke drank with Hobbes. Nicolson's monograph has resurrected a whole school of poetry, the "scientific" poets of the eighteenth century, who spelled out the facts of science in the cadence of enlightened rhyme. The poet James Thomson saw the rainbow through the eyes of Newton:

> *In fair proportion running from the red*
> *To where the violet fades into the sky*
> *Here, awful Newton, the dissolving clouds*
> *Form, fronting on the sun, thy showery prism,*
> *And to the sage-instructed eye unfold*
> *The various twine of light, by thee disclosed*
> *From the white mingled blaze.*

"Spring"

The symbol of the prism posed no threat to the poets of the eighteenth century; it represented instead Enlightenment in all its radiant aspects. But Wordsworth saw the rainbow much as the ancients did and used that sign in the sky as a text for a sermon on natural piety.

I must admit that the cleavage of English science from its art in the nineteenth century has not served to diminish the vigor of either. We might in fact propose that without that sort of differentiation, we would still be back in the world of the eighteenth century. I'm clearly of two minds on this point. I wish there *were* areas of experience—other than pop culture or generational turmoil—on which the two groups of students I had seen that morning could agree. At the Barbican, in that tiresome failure of design, the theater directors were devoting their great skills to producing a cycle of Jean Genet's dramas. On the third floor of the pharmacology building at ancient

Bart's, a sparkling team of scientists, led by Nobel laureate Sir John Vane, was solving the rebus of how sticky blood platelets cause heart attacks.

It is unlikely that modern drama would have differentiated to its absurdist phase, or that we could be influencing heart attacks by antiplatelet drugs, had we not first separated one art from another and one science from the next until no biologist has any idea as to what the astronomers are talking about. We have paid a heavy price for that differentiation. We have made junkyards of our cities, clowns of our rulers, gibberish of our journals, and boors of our chemists. We have made it almost impossible to hold general conversations on topics too complex for Marvellous Mandy and have reduced poetry to a hobby for professors. It is sometimes difficult to be sure what we have gained in the process—other than dramas of Genet and the discovery of how blood clots. I *am* sure that among the advantages we have gained has been a life span long enough for most of us to go expertly about any business we wish, including that of art or science. We have become not one culture, not two, but a thousand. We have become as differentiated as the tissues of our body and for the same purpose: perfection of the life *and* of the work.

Deep under the great trees of Charterhouse Square is the mass grave of 50,000 Londoners who died of the Black Death. High above the ancient grave, across Aldersgate Street, rise the towers of Barbican, where a healthy people of the thousand cultures breathe clean air and Stravinsky pulses among the pylons.

> *I was the Dreamer, they the Dream: I roamed*
> *Delighted through the motley spectacle:*
> *Gowns grave, or gaudy, doctors, students, streets,*
> *Courts, cloisters, flocks of churches, gateways, towers.*

> The Prelude, Book III

We owe to Wordsworth many towers, many walls.

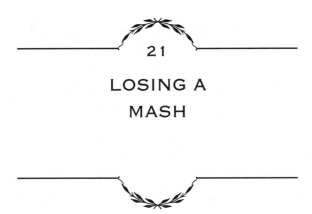

2 1

LOSING A
MASH

We were sitting around a library table on the sixteenth floor of Bellevue Hospital, discussing tactics in the teaching of medicine. Attending physicians, fellows, residents—products of a dozen medical schools over two generations—we had all undergone that intense period of ward training that has been a feature of American medicine since the 1910 report of Abraham Flexner, *Medical Education in the United States and Canada*. We were comparing our own experiences with those described in Melvin Konner's book *Becoming a Doctor*. Konner, an accomplished anthropologist and essayist, went to medical school at the age of thirty-six and proceeded to write an engaging account of his clinical years. His book not only sparked our discussion but earned our admiration. We concluded that what many call the best medical school in Boston must have added to its requirement for graduation the writing of a literate *Bildungsroman*, since no fewer than three other romances of medical education at Harvard in the 1980s have appeared (*Gentle Vengeance*, by Charles Le Baron; *A Not Entirely Benign Procedure*, by Perri Klass; and *Under the Ether Dome*, by Stephen Hoffman). We agreed that the Boston group had gotten the story of medical training just about right, plus or minus local folklore.

Controversy arose over whether clinical medicine couldn't be taught at a more humane pace. Surely, some argued, docs might be more compassionate if they had not been brutalized by the post-Flexnerian hazing ritual, that crunch of learning between the third year of medical school and the first two years on a hospital's house staff: four years without adequate sleep, security, or salary. Laity and medical educators alike have complained that the high-tech doctors who survive this rude ritual have lost touch with the pastoral aspects of medicine; that they lack patience, humanity, and compassion; that they don't listen to their patients.

One of us came to a conclusion remarkably like Konner's: "You put a bunch of bright, competitive people in white coats and scrub suits, assume they've memorized the human genome and the complement cascade, let them write orders on very sick people they've never met in street clothes, and then you wonder why they don't come on like Dr. Schweitzer."

"Nonsense," said another. "Nowhere else in the world is medicine done better. And it's done better here because we *do* teach science, and we *do* train house officers to stay up at night with sick patients who die if someone flubs an order."

I wondered if that attitude wasn't somewhat parochial. Other professions deal with life and death: pilots and politicians sometimes face crises that dwarf those of the clinic. The years of medical training are perhaps no more brutalizing than the apprenticeship undergone by young lawyers, venture capitalists, or brokers. The best of those professionals also work incredibly long hours, have to make equally touchy decisions, and play for very high stakes indeed.

A resident disagreed. "You can say what you like, but in big medical centers like this one, we deal with life and death all the time; we're not just driving an airplane. And when my intern and I drag out of the on-call room at three in the morning we're not doing it for a leveraged buyout or to beat a competitor."

"In fact," observed one of my older colleagues, "that's what we all remember best. Those early mornings in scrub suits with a senior resident yelling at us to put in a line."

"It's like boot camp," said one of the fellows. "You sweat like hell, but you come out a Marine."

An appalled woman resident objected: "That's what scares me, medical school shouldn't turn us into Marines! Soldiers are taught to kill: they're on the wrong side."

I mused out loud. "You know, we may not like it nowadays, but I suppose that house staff training *is* very much like basic training in the military. The Army or Marines may train you to crawl under machine-gun fire, but *we* train you to shock a real heart back to life. The soldier may spend his whole military career buying camshafts, and you might become a dean. But you'll never forget the day you first put on a white jacket, any more than I'll forget the day I put on an Army uniform. Besides, the Army seemed on the right side, then."

It is difficult to remember what turn the discussion next took, because I began to reconsider what I'd just said. Was it in fact necessary to train young doctors in the military mode? Wasn't it possible that we conflated martial with medical training in a puff of imperial pride?

On the office walls of doctors of my generation hang photographs of young men in white uniforms. There, in the front row, is the crew-cut "chief of service," the commanding officer. There, way in back of the fourth row, is that callow chap, the clean-shaven "house officer," who has yet to blunder into middle age. "We were on the house staff together, he was my chief resident," we tell each other. Rank, hierarchy, structure—is that what ought to dictate the teaching of medicine? Or is our military model based on the experience of our teachers? The men who taught us medicine and the men who taught *them* had been to war; many had been the most willing of warriors.

It cannot be said that my own two years in the service left me with undiluted respect for the military model. Not even nostalgia can erase memories of the unspeakable boredom that suffused the peacetime Army in the mid-1950s. During the diluted basic training we received as medical officers at Fort Sam Houston we complained that the Army ran on the principle that the incompetent must lead the unwilling to perform the unnecessary. But there were moments when the pageant got to us, when bugles blew retreat and the colors were furled in a mauve San Antonio sunset that caught the heart. And even the dreary stalemate of the Korean War had not yet diminished some of the Second World War glory of those who trained us. One learned that the ribbons sported by our drillmasters— fellows only a few years older—had been won for chasing the Germans out of Sicily, assaulting the beaches of Normandy, or opening the gates of Dachau. Feelings of gratitude were mingled with constant wonder that this drawling crew of layabouts had beaten the Wehrmacht and kept me from being turned into kitchen soap.

Many of my Army colleagues, snatched from training programs at elite hospitals, muttered their own paraphrases of Proudhon's quip, "I would die for the common man rather than have to live with him for a week." For others of us the service was an eye-opener, a crash course in American diversity, an outdoor seminar in experimental sociology. We were also permitted to lay mines and to fire carbines.

Crawling along the infiltration course at Camp Bullis, or wilting through lectures on the layout of latrines, one hoped eventually to be stationed at the periphery of the new American empire: perhaps near the temples of Kyoto or the forest of Fontainebleau. In the event, I became a turnpike warrior, posted to Fort Dix, New Jersey, so that I could help—in the words of the large Texan who sent us off—"to defend our way of life against the forces of Atheistic Communism that threaten to engulf the Free World."

The charge was easily fulfilled in the pine barrens near Bordentown, where the only Communist I met was a potter from New Hope. I defended democracy for the better part of a year by examining the backsides of recruits and playing pinball in the officers' club. During a rainy fall and dreary winter I volunteered for any assignment that would remove me and my family to more congenial surroundings. It was no small delight, therefore, when I drew by chance temporary duty described by our commanding colonel as "a soft deal on the shores of Lake Ontario." For there, at Oswego, New York, the First Army ran a summer training camp for its antiaircraft artillery units and reserves. The artillery people shot with 90-millimeter cannons at small drone planes, and Army regulations decree that where cannons are fired, a doc is required. I would become the post surgeon of this artillery camp and could move with wife and baby to lovely Oswego on the lake; my duties would be nine to five with weekends free, because the firing range was open only while the technicians from the Cadillac motor division of GM were around to manipulate the radio-controlled target planes.

There was one small complication before we could make what appeared a welcome summer retreat from the heat of southern New Jersey. At the end of May, I would have to move some enlisted men and equipment from Fort Dix to the Oswego range—a few hundred miles to the north—and deliver a mobile army surgical hospital to Camp Drum, which was somewhat farther northeast along the lake, at Watertown, New York. This transfer was routine, the colonel assured me, and would be handled by my executive officer, a Medical Service Corps lieutenant named Hooper. Once Hooper and I had the MASH safely into Watertown, we could take a week off, find homes in the town of Oswego for our families, and bring them up to the cool lake for a summer of fun. I immediately liked Hooper, who hailed from Syracuse. He was in his mid-thirties and had become an optometrist after his infantry days in Europe, only to be recalled during the Korean War,

Author's photo, *Camp Oswego, summer 1956.*

in the course of which he had learned to manage field hospitals. It was left unclear why this amiable man had not risen above the rank of first lieutenant.

Soon enough, I found myself seated next to Hooper in the backseat of a command jeep. We were rolling north at the head of a convoy of eighteen trucks—mainly old Dodge deuce-and-a-halfs—and assorted ambulances. On my lap was a clipboard with a listing of all that the convoy contained, everything from "sterilizer, chrome, portable, twelve gallon vol" to "crutch, wooden, rubber arm support, laminated," all these in quantities that might have seemed excessive for the French retreat from Moscow. In accord with Army custom, I had signed for the whole lot; two carbon copies of my debenture remained at the base. As commanding officer de jure, I was now responsible for millions of dollars of medical and automotive hardware, not to speak of the welfare of the seventy or so enlisted men riding behind us. All seemed in hand, however, since Hooper had exercised his de facto role and gotten the convoy assembled in crackerjack form.

On that fine spring day the sun shone and blossoms of fruit trees dropped like snow on the hood of our jeep. Hooper and I wore dress khaki with spiffy combat boots; the men wore green fatigues with shiny brass. On their arms, many sported red crosses on white bands to distinguish medics from the drivers; small pennants of red and white fluttered from flagstaffs on our fenders. The scene was very martial and perhaps for that reason its captain led this unit of the First Army through New Jersey whistling the refrain from "Red River Valley." In my ears the lyrics were those of the Abraham Lincoln Brigade: "There's a valley in Spain called Jarama . . ."

We left Fort Dix at seven on a Saturday morning and were not due in Watertown—Camp Drum—until Monday morning at eight. Plans were to make it to Oswego in one long trek and to drop a few men and supplies for the artillery range there. After two nights at Oswego, we would move the MASH to Camp Drum bright and early on Monday morning. Two disasters interrupted this plan. In the first instance, several of the trucks developed mechanical failure, and we were very slow at getting them going again. We were in consequence behind schedule when we arrived in Glens Falls late on Saturday afternoon, to telephone ahead to Oswego. The second foul-up was not our fault. The artillery range, commanded at the time by a colonel whose interest in 90-millimeter guns was subordinate to that in 90-proof gin, had made arrangements for two officers, but not for a MASH. It became clear that no tents were ready for the men and no victuals were available;

the training season was not to begin for another three weeks. We ourselves were not yet under Oswego orders and were in any case not supposed to reappear there until after we had signed over the MASH to Watertown. We were on our own. It was clearly impossible to make it to Camp Drum that day, so Hooper and I were left with the problem of providing housing and food for three-and-a-half score men and the not simple problem of over-night parking for a MASH.

Hooper made what I considered a brilliant suggestion. "Look, Captain, all the enlisted men are from the First Army area, most of them are reservists, from the Troy–Utica area. They all come from New York, or at farthest from western Massachusetts. Tell them you'll give them the rest of the weekend off. They'll go home, eat, sleep, and get to Camp Drum on their own by Monday. That way we don't have to feed or house them. All they have to do is fill up their trucks with gas."

I wondered about the logistics of this for a bit, but then Hooper turned the tide of doubt. He had learned that many a young doctor's sense of duty could be diverted by an appeal to the table. "Not only can the men get a good night's sleep, but I know the Finger Lakes region well. There's this fabulous restaurant in Skaneateles—Krebs', it's called—and they have some rooms. They've got this unbelievable New York white wine you won't find anywhere else in the world, and then they serve blue trout in pure butter."

What was good enough for Faust was good enough for me. Mephisto Hooper bought my soul for the price of blue trout and white wine. I decided forthwith not to check with the CO at Watertown but to take action on my own. I spelled out the bargain to the men of our MASH: we would all assemble at the gates of Camp Drum at eight on Monday morning, they were free to head for the sacks at home on condition they gave us their word to return promptly, with full gas tanks. They seemed grateful and happy, and dispersed within minutes to drive ambulances and deuce-and-a-halfs over the highways to Troy, Utica, and points east.

With Hooper now at the wheel, we sped on to Krebs'. I wish that I could report that the two dinners we had there were worth all that followed, but if truth be told, I remember chiefly the wine. The menu was ample but plain; the fare was distinguished with respect to its ingredients but not its prepa-ration, which merely provided a substrate for the delicate flavor of the then undervalued wines of the Finger Lakes. But while perhaps not then worthy of a Michelin star, the inn was more than adequate for two off-duty officers;

we spent the time between meals playing cards. Our *grande bouffé* ended, we were able to arrive at Watertown at seven on Monday morning by dint of driving through much of New York State in the middle of the night. We parked our jeep about two hundred feet from the gates of the camp and made our morning ablutions with the aid of helmet liners and water from a canteen. Spit-and-polish clean, Hooper and I waited for the troops to arrive.

By five minutes before eight, I had become worried and impatient; by eight I was in agony. None of our vehicles had arrived. We watched as car after car filled with troops from the camp rolled through the gates. Inside, the place was a bedlam of truck noise, troop echoes, bugle calls. Outside the gate, not an army truck was to be seen. Eight-fifteen. I was wet with angst. I was going to be court-martialed for having lost all those trucks, all those millions of dollars; I would be in debt the rest of my life. I looked at the clipboard and could not believe that I had ever signed my name to such a manifold. A "scissor, chrome, surgical, 3 1/2 inches" cost $14.75—and there were ten gross of those!

I could not bear to speak with Hooper. He reassured me that the boys would be along any minute and that no one expected a convoy to be exactly on time. I worried some more. It was now nine and we were still alone in our jeep. I told Hooper to wait by the gate and took the jeep into camp. The sentry responded with a salute; I identified myself and gave him a copy of our orders. He asked where the rest of my unit was, and I stammered that they were "forming up down the road"; he directed me to post headquarters.

They were expecting me at the colonel's office, and I was shown in immediately. It was now nine-fifteen. The colonel, whom I'll call Cooper, because he looked like Gary, was an infantryman. His chest displayed battle ribbons of two wars and featured purple hearts and bronze stars sufficient for the whole cast of *Sergeant York*.

"Captain Weissmann, reporting, sir!" I gave him my best copy of a Hollywood salute.

"Y'all bring that rollin' hospital in heah?" he asked me in an accent I will not attempt any further to reproduce.

"Well, sir, not exactly. I mean, I am here, and my exec is here. We are, actually, ah, er, waiting for the men to form up."

He rose from behind his desk, went to the window, and saw only my now empty jeep on the empty roadway. He picked up a copy of my orders, accepted the clipboard from my hand, looked them both over for half a year, and finally gave me a hard stare.

"Weissmann, that's a New York name, isn't it?"

I had a hunch what he meant by that. "Yes, sir."

"You were due in at eight o'clock in the morning with eighteen deuce-and-a-halfs, some ambulances, and seventy-odd men. A whole effing United effing States effing Mobile effing Army . . . Where in hell are they?"

Since I had neither gasoline nor match at hand I was unable to commit the first act of suicide that sprang to mind. American reserve Medical Corps officers are not issued swords; that ruled out the second. I ventured the third by opening my mouth. I told him about the bargain I had struck with the men.

"You're supposed to be their commander, not their camp buddy! I don't care if you are some smart-ass New York doctor, you can't just let a bunch of men go off in government property all over the state. You'll be in hock right up to your Park Avenue ears! If I don't hear those trucks rolling in soon, you're liable to be in the stockade mighty quick and mighty poor. Before you scatter a bunch of men and trucks all over the map, boy, you ask someone!"

It was now closer to ten and the only sound I could hear was the firing squad in the next room checking their carbines. I wondered who would clean up the puddle of sweat I was leaving on the varnished floor.

The colonel took one look over the empty street, then he sat down facing me. "You really screwed up, didn't you? Well, there's only one thing for me to do now." He reached into his desk drawer.

He's going to kill me right now with his service revolver! I thought. He brought out a fairly full bottle of Jack Daniel's, followed by two bar glasses. "Doctor, why don't you and I just sip some whiskey while we wait for the boys to come in off their holiday?"

And so, for the next hour and a half, my petrified, younger self and the tall man from Tennessee sat there as slowly, very slowly, the street outside filled up with trucks and ambulances and the men of our MASH were hustled into rank by a very sheepish Hooper.

I thought about that morning at Camp Drum as our discussion at Bellevue came back into focus. It had covered the map of clinical learning and we wound up by defining its limits as "drill" and "problem solving." We decided that the discomfort we feel nowadays in the presence of the military ritual extends to the military aspect of a doctor's house staff and student drill, the rite of passage we had all experienced. But we also agreed that university teaching differs from military teaching in one important way; whereas in the absence of drill battles may be lost, without problem solving, the war is over.

Questions remained. To what extent should doctors train by rote, to what extent by trial and error? When the stakes are as high as life and death, is it appropriate to work by the old rule of "See one, do one, teach one"? Are we acting responsibly when we permit the chance of error by a doctor in training? Who pays for each goof? Is there a contradiction between good medical care and learning by doing?

From Thomas Arnold at Rugby to John Dewey at Columbia, educators have spelled out the dialectic between the *form* of drill and the *function* of problem solving. We might argue that, as in the canon of architecture, medicine is best taught when form follows function.

No one was hurt by my loss of the MASH. There was no battle missed or opportunity wasted. No one died and no one suffered. Except for my small embarrassment, no sentiments were greatly bruised. Hooper and I emerged as more dutiful soldiers in consequence of the colonel's largesse. We learned the strength of the rules by having been forgiven their violation. We had been caught by that oldest of tricks by which the young are taught: a pardon. It is a difficult lesson for a prospective teacher to ignore.

When the convoy finally formed up and drove down the parade strip to our bivouac area, it made quite a fine sight in the cool northern spring sun. Our jeep in the lead, we passed between ranks of new 90-millimeter guns waiting on their beds to be tracked to the range. Their crews gave us a wave. I was reminded of a passage by George Orwell in *Homage to Catalonia:*

> As our train drew into the station a troop-train full of men from the International Column was drawing out, and a knot of people on the bridge were waving to them. It was a very long train, packed to bursting-point with men, with field-guns lashed on the open trucks and more men clustering round the guns.... It was like an allegorical picture of war; the trainload of fresh men gliding proudly up the line, the maimed men sliding slowly down, and all the while the guns on the open trucks making one's heart leap as guns always do, and reviving that pernicious feeling, so difficult to get rid of, that war *is* glorious after all.

Difficult, said Orwell, meaning not impossible.

22

TO THE NOBSKA
LIGHTHOUSE

The Nobska lighthouse stands guard over shoals at Woods Hole where the Atlantic runs between Vineyard Sound and Buzzards Bay. The bluff on which the lighthouse sits is the highest on that stretch of the coast and on fair days yields splendid views of upper Cape Cod and its islands. The prospect is never fairer from Nobska point than on early mornings in October, when—in the words of Justice Oliver Wendell Holmes—"the wind blows from the west and the air is clear." Recently, such a morning found me running uphill to the Nobska lighthouse. A plangent sea was on my left, thickets of wild grape and bittersweet were on my right. In the distance across the sound, the highlands of Martha's Vineyard flashed pink in the dawn. Nearer by, the hillocks of Naushon Island flickered mauve and green. A few cirrus clouds were backlit on the eastern horizon, the sky above was clear. On the crest of the hill the westerly breeze rustled low conifers and floated a strike force of gulls. Breathless, I stopped at the top of the bluff, convinced that the shores of Arcadia could offer nothing so bracing as this New England sunrise.

The Nobska light station is placed on a tidy promontory seventy-six feet above sea level. Its lawns are closely mowed and its hedges tightly clipped, its buildings display the prudent ordinates of Yankee architecture. The horizontal axis is defined by the keeper's house: twin, gabled saltboxes joined into a single cottage. Modest vertical accents are provided by three chimneys, a steel radio antenna, and a signal tower from which small-craft warnings flutter on stormy days. The chief vertical axis is announced by the monumental shaft of the lighthouse, a tapered tower some forty feet high made of whitewashed steel-encased brick. The tower is topped by a black lacework cupola in which the light source is housed behind a ten-sided cage of window panes. A single 150-watt bulb refracted by a cunning Fresnel lens generates 7,000 candlepower and can be seen for more than ten miles.

With stars and stripes fluttering from a freestanding flagpole, with deep shadows cast by morning light under the cottage eaves, the scene was an Edward Hopper canvas brought to life. Other themes sprang to mind. I had read that the lens of the lighthouse had been ground in 1828 by New England craftsmen from designs by Augustin-Jean Fresnel (1788–1827). Fresnel was a Jansenist from the Vendée who in the post-Napoleonic restoration pursued his optical theories while on tours of duty with the civil service. He was convinced that light was a wave rather than a particle, as Newton had taught. The young Frenchman arrived at his equations in settings fit for making waves of light: the Lighthouse Commission and the *Corps des Ponts et Chausées* (Bridge and Road Corps). His more practical work resulted in the replacement of mirrors by lenses in lighthouses the world over. Fresnel's early death from tuberculosis ended a blossoming reputation not only in physics but also in moral philosophy; he had become a keen apologist for the puritanical doctrines of his sect. He might have been pleased to know, were he among the elect, that for more than a century and a half his lens had cast light over waters that the stern Pilgrims had mastered. He might also have subscribed to the closing sentiments of "Fair Harvard":

> Let not moss-covered error moor thee by its side,
> While the world on truth's current glides by;
> Be the herald of light and the bearer of love,
> 'Til the stock of the Puritans die.

Reasoning that I had been diverted from matters Arcadian to refrains academic by the anoxia of uphill running, I started downhill with relief. Arcadia had waited. Nobska beach, in the cove below the lighthouse, was at low tide. Its sands reflected the morning sun and the breeze played with tendrils of seaweed at the tidemark. By the roadway, bracken on the low dunes glowed in autumn paisley. The beach was empty, the only artifact visible an old Corvair parked some thirty yards away. The single object in motion was on the water: the first ferry to Vineyard Haven had rounded the point of Woods Hole harbor. It made a dashing sight, with its white hull, black smoke, and frothy wake against a turquoise sea. Exhilarated, I kept running along the short length of the beach back to the center of town.

I had almost left the Corvair behind me when I noticed a striking bumper sticker placed prominently on the flivver's rear. It showed the abortion rights symbol, a black coathanger on which the red international logo for "prohibited" had been superimposed. NEVER AGAIN was the legend, and NARAL (the National Abortion Rights Action League) was noted as the sponsor. The car was empty and bore no other identification except its Massachusetts plate. At that moment, I spied what must have been the owner of the car emerging from behind a bend of the cove. At that distance I could see she was a fit, middle-aged woman with short gray hair, dressed in khaki shorts and a blue denim work shirt. She was busy gathering seashells: stopping, stooping, starting, she looked like a busy shore bird. Suddenly she halted and, shielding her eyes, looked straight out at the sea and the passing ferry. Her trim form echoed the vertical of the lighthouse above; the sun caught her face and was reflected by the galvanized tin bucket she had put down beside her. Solitary against the beach, her Keds in the sand, she could have served, I thought, as a model for a low-keyed statue of Reason.

The woman in her Keds, the NARAL logo, the lighthouse—the Enlightenment?—spurred associations that turned the rest of my outing into a rumination on the days of the rusty coathanger. My mind was, literally, jogged back to Bellevue Hospital in 1959 and the era when abortion was illegal. Coathangers, rusty or not, were not the chief instruments of botched home abortions. The victims treated by my generation of house officers had been invaded by knitting needles, rat-tailed combs, and—sad to say—the metal probes used by plumbers. It is not true that only poor women bore the brunt of injury; fear of parents, fear of their partners, and fear of professional abortionists often brought daughters of the middle class to the emergency room. Rich and poor alike were at risk for the major complications of all that clumsy instrumentation: bleeding and sepsis, frequently both. For reasons not hard to imagine, the patients arrived most often at midnight and too often in shock.

Sometime after midnight on a warm Labor Day weekend thirty years ago, the Keds that I remember to this day were neatly stowed side by side under the bed of a young woman who lay febrile, breathless, and barely conscious in the emergency ward of Bellevue. One white sneaker was immaculate, the other was stained by two splotches of blood that had soaked into the canvas uppers and dried on the rubber of the instep. Most of what I recall of the larger scene was also in primary colors. The three doctors and two nurses at the bedside were in white, as were the curtains around the bed, the sheets, and the patient's gown. The blankets alone were gray, city issue, but the young woman had flushed, freckled cheeks and flame-red hair. Her lips were blue and her temperature was 104. Trying to rouse her, the nurse called her Kate; I have forgotten her last name. She was a nineteen-year-old nursing student from the school across the street and had been brought in after a fumbled attempt at self-abortion. I was the chief medical resident of Bellevue Hospital and all of ten years older than the patient. I ordered some epinephrine, diluted it, and injected a tenth of a cc in her skin.

"We're going to look for the Thomas lesion," I said, as I pointed to the pale forearm taped to a plank with white gauze. "If there is endotoxin on board, that little spot should be necrotic in a few hours." First white, then red, then black; color and then the absence of color.

The human details of the case are less clear in memory than the stark tableau of the emergency room. I seem to recall that Kate was a farm kid from Dutchess County. She had missed a period, turned positive in a furtive rabbit test of the day, and tried to instrument herself with the help of a friend late on Saturday night. The procedure would have involved a hand mirror, some stolen local anesthetic, and a steel knitting needle sterilized over a candle. But it hadn't worked, all that happened was a cervical tear and diffuse bleeding, which the kids tried to stop with gauze. Shaken and afraid, Kate had gone to ground in the student nurses' dorm all day Sunday. By Sunday night she was again oozing large amounts of blood and had developed shaking chills. She was brought to the receiving area of Bellevue by her friend, who also provided all the details of Kate's history. By the time I saw her early on Labor Day morning, she should have been better. The gynecologists had typed, cross-matched, and transfused three pints of blood, performed a dilation-and-curettage, and sutured her cervical wounds to stop the bleeding. But complications arose. Her temperature climbed, her white cell and platelet counts dropped, bleeding started again from every needle and suture site. Suspecting that their patient was bleeding from disseminated intravascular coagulation due to gram-negative sepsis—a disorder of blood clotting caused by bacterial endotoxin—the surgeons paged the medical resident on call. He in turn knew that I was in the house, that I was working with bacterial endotoxin, and that I was up to date on gram-negative shock. I had been briefed by an expert, Lewis Thomas, our chief of service.

So there we were, like doctors before us and after, up to date but quite at sea, at the bedside of a sick young woman, searching in the only way known at the time for endotoxin in blood. We gave her more blood transfusions, intravenous broad-spectrum antibiotics (Aureomycin? Chloromycetin?), and prayed that her kidneys wouldn't shut down. I forget now what bug it was that grew out of her blood cultures the next afternoon, *Aerobacter* or *Pseudomonas*. I do remember that by then things could not have gotten worse. Her kidneys had failed, she never brought her oxygen up, she kept bleeding. We could not pull her out of shock and she died before we found out whether the Thomas lesion had turned positive.

I also remember meeting her parents the next day in the squalid waiting room of the old hospital. A postmortem examination by the Office of the Chief Medical Examiner, which was mandatory in such cases, had shown that she had died of septic shock with bilateral renal cortical necrosis and pulmonary edema with blood in the lungs. The parents appeared grief-stricken but not entirely surprised by the news that their daughter had died from a fumbled abortion. They were stern fundamentalists, American Gothics, and deep in their rustic hearts they seemed to have expected that sin would catch up with Kate, who was the "fastest" of their three daughters. She had broken their hearts when she had run away to the big city to become a nurse; the profession in those days was tainted by a touch of the profane in the minds of small-town folk. As we spoke of their daughter's death, they looked at me as coldly as if I myself had raped and killed her. In cool fury, directed at cities, hospitals, doctors, nurses, women—who knows what—the father permitted himself to say, "God punished her. She must have deserved it. She's better off this way." He took his wife's hand, asked the way to the mortuary, and left the building. Through the glass-paneled door I saw him stop, carefully fit a black felt hat to his pate, and proceed up East Twenty-sixth Street with his wife in tow.

Absorbed in those ancient recollections on my morning ramble I almost failed to notice that I had come abreast of another Woods Hole landmark, the seaside Church of the Messiah. Its new copper gutters shone in the glory of fall. Morning light also caught the salmon and gray of its cemetery stones; the green lawns had not yet been cleared of overnight leaves. I reckoned that poor Kate had been born a generation too soon on two counts: these days abortion is legal, and we know much more about the cause and treatment of endotoxin shock. Young women like Kate need not die. She would have been forty-nine by now; I imagined her standing under the

Nobska light, her Keds in the sand, her Corvair parked down the road. More power to NARAL. But as I looked up at the New England steeple, it struck me that the Puritan fathers would have been as stern on abortion as Kate's father. I'm not sure how they would feel about endotoxin.

The meliorist path of our secular republic is not the road mapped out by the elders of Plymouth. The Yankee patriarchs found it easy in the name of their God to blame His victims and easier still if their names be women. Dr. Cotton Mather would have been pleased by President George Bush's veto of a bill permitting the use of federal funds to pay for abortions of indigent women who have been the victims of rape or incest. Governor Bradford of Plymouth Colony might have gone a touch further. On September 8, 1642, some thirty miles north of the Church of the Messiah, a sixteen-year-old youth named Thomas Granger was convicted of unnatural sexual acts and executed by order of the magistrates of Plymouth. According to Governor Bradford's journal of the event, *Of Plymouth Plantation 1620–1647*, the lad had been

> detected of buggery, and indicted for the same, with a mare, a cow, two goats, five sheep, two calves and a turkey. . . . A very sad spectacle it was. For first the mare and then the cow and the rest of the lesser cattle were killed before his face, according to the law, Leviticus xx.15; and then he himself was executed. The cattle were all cast into a great and large pit that was digged for the purpose of them, and no use made of any part of them.

No mention is specifically made of the turkey, but Bradford tells us that the elders were worried that innocent sheep might be slaughtered. They forced young Granger not only to confess, but to identify his former playmates: "And whereas some of the sheep could not so well be known by his description of them, others with them were brought before him and he declared which they were and which were not."

The Pilgrims' attention to details of sexual conduct was coupled to a strict regard for biblical authority. Shortly before Granger's execution, Bradford asked three local divines to find legal and scriptural precedents for the death penalty in cases of sexual deviance. He asked several specific questions related to offenses of sex and received appropriate answers in which no graphic detail of plumbing was omitted. As one might expect from a future president of Harvard, Charles Chauncy's reply had more citations in Latin and English than the other two responses, and the most from the Old Testament. His answer was also the strongest in rhetoric, the freest of sentiment, and the sternest in tone.

The Answer of Mr. Charles Chauncy

An contactus et fricatio usque ad seminis effusionem sine penetratione corporis sit sodomia morte plectenda?

> Question: The question is, What sodomitical acts are to be punished with death, and what very fact committed (*ipso facto*) is worthy of death, or if the fact itself be not capital, what circumstances concurring may make it capital? The same question may be asked of rape, incest, bestiality, unnatural sins, presumptuous sins. These be the words of the first question. The answer unto this I will Lay down (as God shall direct by His Word and Spirit) in these following conclusions.

Chauncy answered that the Mosaic laws are "immutable and perpetual" and grounded on the law of nature, indeed, that all the sins enumerated are punishable by death. Quoting extensively from Luther, Melanchthon, Calvin, and other fathers of the Reformation, he reassured the governor that

> Then we may reason . . . what grievous sin in the sight of God it is, by the instigation of burning lusts, set on fire of hell, to proceed to *contactum et fricationem ad emissionem seminis*, etc. and that *contra naturam*, or to attempt the gross acts of unnatural filthiness. Again, if that unnatural lusts of men with men, or woman with woman, or either with beasts be to be punished with death, than *a pari* natural lusts of men toward children under age are so to be punished.

These themes, and their canonic variation, were also sounded by the two other elders of the Church, Mr. Rayner and Mr. Partridge. The Old Testament called for the death penalty for "unnatural vices" or offenses to God, and the Puritan preachers found ample precedent in Mosaic law for divine retribution. But a closer reading of Charles Chauncy's reply to Governor Bradford yields a remarkable passage, which must contain the first—and most severe—American argument for the prohibition of abortion. The "pro life" movement may be said to have begun in 1642; it was announced in concert with a call for death:

> In concluding punishments from the judicial law of Moses that is perpetual, we must often proceed by analogical proportion and interpretation, as *a paribus similibus, minore ad majus*, etc.; for there will still

fall out some cases, in every commonwealth, which are not in so many
words extant in Holy Writ, yet the substance of the matter in every kin
(I conceive under correction) may be drawn and concluded out of the
Scripture by good consequence of an equivalent nature. *As, for example,
there is no express law against destroying conception in the womb by potions,
yet by analogy with Exodus xxi.22, 23, we may reason that life is to be given
for life* [my italics].

Perhaps Chauncy's doctrine that life is to be given for a life found its highest
expression in the era of the rusty coathanger. Social historians have traced
some of the prohibitions against abortion in the days before *Roe v. Wade* to
the severe Mosaic laws of Plymouth Plantation. Puritan values and Catholic
teaching made Massachusetts the last state in the Union to permit doctors
to prescribe birth control for married women (1966). The prohibition of
unpopular private behavior has a long history in New England; not only
sodomized cattle have been sacrificed in Massachusetts. The Puritans'
preoccupation with sexual offenses, their obsession with anal coitus and
seminal emissions, their fear of filthiness and the unnatural have set the
darker themes of political and religious discourse in America for more than
three centuries. H. L. Mencken defined Puritanism as "the haunting fear
that someone, somewhere might be happy." A more dynamic interpretation
of the excess attention paid to pudenda by the Pilgrims might explain why
the life that was to be given for a life was invariably a woman's.

More agreeable aspects of the Puritan legacy came to mind as I loped
the last mile home. My path took me to the back of Eel Pond, a natural
marina around which the village is disposed. The last yachts of the season
and assorted skiffs were berthed in the small harbor; its periphery was
ablaze with autumn elm and maple. Cormorants ruffled the pond. Across
the water, the sun shone on the colonial cupola of the Woods Hole Oceanographic Institute (WHOI). A flag snapped in the breeze above the pediments
of the Marine Biological Laboratory; the early sun was reflected from
windows of its library. The hum of machinery which carried softly from the
labs reminded me that both of those scientific institutions are in their
halcyon days. That summer, deep-water submersibles from WHOI had
found the battleship *Bismarck* at the bottom of the Atlantic, while with the
spiffiest of new microscopes scientists of the MBL had discovered the
molecular motors of mitosis. The morning panorama of a maritime campus,
neat and shipshape, already busy at work, was a hard-edged illustration of
Puritan values. Bradford and Winthrop would have approved: their land

still housed heralds of light and bearers of love; truth's current had not passed Woods Hole by.

Governor Winthrop expressed a millenarian vision of the New Jerusalem when he spoke of the colony "as a City upon a Hill, the eyes of all people are upon us." The eyes of all people were no less important to the elders than the ears, for the Puritan leaders were preachers of the Word. But above all, they were men who worshipped ideas. "Every man makes his God," wrote Dr. Oliver Wendell Holmes to Harriet Beecher Stowe; "the South Sea Islander makes him out of wood, the Christian New Englander out of ideas." Holmes—son of a Cambridge minister, professor of anatomy at Harvard, and father of the Supreme Court justice—knew that he and his kind, the sons of the Puritans, could never "get the iron of Calvin out of our souls." The other sentiment he could not erase was a Calvinist sense of the elect, of belonging to an elite whose sainthood was made visible by the products of mind.

Historians of the Puritan revolution teach us that a ministry of educated men was required in order to replace the sacramental priesthood of the Roman Church. Education based on the Word of the Bible, and not the authority of a church, was expected to assure the victory of Puritan mind over papal matter. What was less expected was that biblical education, especially in the New World, would also lead to a democratic, individual response to society: all this and the rise of the middle class. Friedrich Engels, writing in 1892, pointed out that Calvin's creed was one fit for the most advanced bourgeoisie of the Puritans' time and that Calvin's constitution was "thoroughly democratic and republican." Almost a century later, the historian Michael Walzer has described the Puritan clergy in England and America as "educated (or self-educated) and aggressive men who wanted a voice in church government, who wanted a church, in effect open to talent. . . . Decisions would be made by prolonged discussion and natural criticism, and finally by a show of hands. Somber, undecorated clothing would suggest the supremacy of the mind. . . . The Puritan ministers provide perhaps the first example of 'advanced' intellectuals in a traditional society. . . . Its first manifestation was the evasion of traditional authority and routine."

A disproportionate number of these advanced intellectuals came to New England; between 1629 and 1640 some one hundred Cambridge men arrived in America. Charles Chauncy would have returned to a chair by the Cam, in England, had not the new Cambridge made him its master. One suspects that Chauncy, like other intellectuals since, may have evaded traditional authority and routine in order to impose an authority and routine of his own.

Traditional authority and routine do not rule Woods Hole today. From the Nobska lighthouse to Eel Pond, moss-covered error keeps no one at its side; the Yankee landscape and the learned institutions throbbing by its sea bear witness to the nobler side of the Puritan effort. From the Charles to the Housatonic, New England considers itself, perhaps rightly, as the intellectual arsenal of our democracy. The darker side of the Puritan endeavor—patriarchal, bigoted, severe—seems for the moment to be under wraps. Indeed, even the history of Puritan terror, the tradition of Chauncy, Cotton Mather, *et al.*, are in good hands these days. Historian Gordon Wood, writing in *The New York Review of Books*, contends that the two leading scholars of Puritan life and letters, successors to such Puritan stock as Perry Miller and Samuel Eliot Morison, are Sacvan Bercovitch (Harvard) and Andrew Delbanco (Columbia). From Bercovitch and Delbanco one gathers that the Puritan terror was at least in part a response to the alienating experience of absolute power in the face of absolute wilderness. Those scholars remind us that the men who hanged witches in Salem also founded Harvard. Nowadays that university is neither bigoted nor severe. It remains a place for the worship of ideas, and in that sense its scholars are all sons of the Puritans.

A future historian may judge that our era, with its free-and-easy social arrangements, our uncommitted youth, casual sex, foul manners, terrifying streets, our infatuation with the gaudy, the rich, the drugged, the besotted, the violent—our culture of Lawrence Taylor, Mick Jagger, and Donald Trump—is not a happier place than the strict New England of 1642. Young men have died for sodomy in both societies, young women have been murdered then and now. If the traditions of Chauncy and company were distantly responsible for the death of Kate in the Bellevue of 1959, one might ask what religious or secular orthodoxy is responsible for deaths at Bellevue in 1990, for the victims of rape, murder, addiction, and AIDS?

Looking at the peaceful scenes of New England, I thought of the new Bellevue, where I work most of the year. An unshaven derelict named Steven Smith, discharged after "treatment" for violent behavior, "dressed like a doctor" in a scruffy scrub suit, put a stethoscope around his neck and roamed around the hospital at will. On Saturday night, January 7, 1989, he raped and murdered a young pathologist who was working late in her office. I was reminded at the time of the death of young Kate thirty years earlier: a life for a life. Kathryn Hinnant was also pregnant. For whose life did she pay? Steven Smith's? That same future historian might find that we are unknowing accomplices in Kathryn Hinnant's death as Charles Chauncy was in the execution of Thomas Granger. He might, if conservative, argue that Dr. Hinnant died for our liberal creed. He might be able to find evidence that our closed asylums, our practice of permitting the mad and violent to prowl the streets, had results more lethal than the Puritan terror.

Saturday nights in Bellevue Hospital were less hazardous for young doctors in my day. The chief resident, like all house officers, lived in the house staff dormitory; if lucky, one saw one's family on alternate evenings. This arrangement gave one a goodly amount of time in the hospital and made it possible not only to follow patients but also to do simple research. In the summer and fall of 1959 one of my projects was to study how epinephrine caused necrosis of skin in rabbits treated with endotoxin. Lewis Thomas had established a small animal room on the sixth floor of the old Bellevue, and there he put me to work measuring fever in rabbits—another bioassay for endotoxin. The procedure involved the frequent insertion of small thermosensitive probes into the rectums of bunnies; I am persuaded that my postdocs who do density gradient experiments with clean gloves have a better deal these days.

In the event, the epinephrine project was part of an effort to find out the mechanism of the Shwartzman phenomenon, the thorough exploration of which had made my mentor's scientific reputation. Here is his account of the *local* phenomenon in *The Youngest Science*:

> A small quantity of endotoxin is injected into the abdominal skin of a rabbit, not enough to make the animal sick, but just enough to cause mild, localized inflammation at the infected site, a pink area the size of a quarter. If nothing else is done the inflammation subsides and vanishes after a day. But if you wait about eighteen hours after the skin injection, and then inject a small non-toxic dose of endotoxin into one of the rabbit's ear veins, something fantastic happens: within the next two hours, small, pinpoint areas of bleeding appear in the prepared skin, and these enlarge and coalesce until the whole area, the size of a silver dollar, is converted into a solid mass of deep-blue hemorrhage and necrosis.

Kate, the nursing student, had died of the *generalized* Shwartzman phenomenon, in which two appropriately timed injections of bacterial endotoxin produce bilateral renal cortical necrosis, pulmonary edema, and—too frequently—death. Whether local or generalized, the Shwartzman phenomenon leads to the clumping of platelets and especially neutrophils within the circulation. These tend to become sequestered in the small capillaries of kidney and lung or to attach to the sticky walls of blood vessels of the prepared skin site. By 1959 Thomas and coworkers knew that if white cells were removed from the equation, the lesions did not develop. A few years before, he had discovered that small amounts of adrenaline—epinephrine—which alone had little effect, would cause hemorrhagic necrosis in the skin of rabbits prepared with a previous injection of endotoxin. His younger disciples, and soon the whole house staff of Bellevue, came to call this lesion "the Thomas test" since we applied it to humans to judge whether they had endotoxin in the circulation. Ironically, on the night Kate was admitted I had returned to Bellevue to check on the skin of some rabbits whose "Thomas lesion" had been abrogated by endotoxin tolerance.

Nowadays, thanks to a generation of investigators, but especially to Drs. Timothy Springer, Ramzi Cotran, Michael Gimbrone, and Michael Bevilacqua of the Harvard Medical School, we have a pretty good notion of why exactly the Shwartzman phenomenon comes about. Indeed, we know so many of the proteins and genes involved that we are at the point of dotting the *i*s in iC3b and crossing the *t*s of its TATA box (these are abbreviations for the relevant molecules). When endotoxin is injected, the walls of blood vessels (endothelial cells) are made receptive and sticky, because the endothelial cells display specific adhesive molecules for white cells. In turn, white cells display specific adhesive molecules which permit them to stick to each other and to blood vessel walls. In the process, molecules such as interleukin 1, tumor necrosis factor, and complement split products are let loose in the circulation, with consequences described earlier in these pages. Those gymnastics of cell regulation produce the effects that Lewis Thomas knew by 1959:

> In the Shwartzman phenomenon, cell death is caused by a shutting off of the blood supply to the target tissue. After the second injection, the small veins and capillaries in the prepared skin area become plugged by dense masses of blood platelets and white cells, all stuck to each other and to the lining of the vessels; behind these clumped cells the blood clots, and the tissue dies of a sort of strangulation. Then the blood vessels suddenly dilate, the plugs move away into the larger veins just ahead, the walls of the necrotic capillaries burst, and the tissue is filled up, engorged by the hemorrhage.

We never did figure out how epinephrine worked. But when, at the end of my Woods Hole run, I thought about poor young Kate and her botched abortion that Labor Day weekend in 1959, I had an idea for an experiment. What if endotoxin were to increase the number of epinephrine receptors on the surface of endothelial cells, as it increases the number of adhesive molecules? What if that poor young woman died because the blood vessels of her kidneys closed in an overeager response to epinephrine? What if . . .?

Looking at the gracious landscape of the Puritans, I realized that the pursuit of ideas in that university seventy miles to the north—by Bevilacqua, by Holmes, and yes, by Chauncy—had provided the facts I needed to ask those questions. Not only Fresnel lenses project light. I also realized, of course, that I was looking for a scientific solution to a social problem. The Kates of today need not die, because abortion is still legal, and as a doctor I strongly resent those movements that would permit return of the rusty coathanger. I have no idea when life begins, but I am sure when it ends. Absent a new Puritan terror, young women—children, indeed—will become pregnant. Absent recourse to safe abortion, some of those, and mainly the poor, will die in blood and pain. Liberal programs, when they go wrong, may disrupt social *order*. But the error that fundamentalists make is to value creed over social *justice*. Mosaic laws are laws for old men. In the slogans of the campaign against legal abortion, I hear echoes from the Plymouth of 1642: A life for a life is a call for revenge by elders on the bodies of young women.

In advocating the cause of keeping abortion legal and safe, I am drawn to the finer side of the Puritan temper. Dr. Oliver Wendell Holmes, after an apprenticeship in Paris, brought home not only a French microscope but also the enlightened Gallic habit of clinical investigation in urban hospitals. Discovering that young Boston mothers died of puerperal fever because doctors carried the infection from bed to bed, he defined the problem, described the obvious solution, and was roundly denounced by the traditionalists. In 1855 Holmes replied to his elders on behalf of the young women of Massachusetts much as I would speak on behalf of those who will surely die if *Roe v. Wade* is further diluted: "I am too much in earnest for either humility or vanity, but I do entreat those who hold the keys of life and death to listen to me also for this once. I ask no personal favor; but I beg to be heard in behalf of women whose lives are at stake, until some stronger voice shall plead for them."

23

TITANIC AND
LEVIATHAN

*Why upon your first voyage as a passenger, did you yourself feel such a
mystical vibration, when first told that you and your ship were now out
of sight of land? Why did the old Persians hold the sea holy? Why did
the Greeks give it a separate deity, and own brother to Jove? Surely this
is not without meaning.*

HERMAN MELVILLE, *Moby-Dick*

We had almost forgotten that *Atlantis II* was returning to Woods Hole that
morning when the buzz of helicopters overhead reminded us. Woods Hole
is but one of several villages in the township of Falmouth, on Cape Cod,
and the whole seaside community was expected to turn out. From the
windows of the laboratory we could see the ship approaching less than a
mile offshore. The July sky was cobalt, the sea a Prussian blue, and the sun
sparkled on whitecaps in Vineyard Sound. It was nine-thirty on the clearest
morning of summer. The research vessel was headed home in triumph after
its second voyage to the wreck of the *Titanic*. Above, the ship was circled by
a corona of helicopters and photo planes; on the water, a flotilla of power-
boats and racing sloops kept pace.

In shorts and lab coats we rushed down the stairs to cross Water Street
in order to be on the WHOI dock when the *Atlantis II* pulled in at ten.
"WHOI" is the acronym for the Woods Hole Oceanographic Institution, the
youngest of the three scientific installations that share the harbors of our
small village. The others are the Marine Biological Laboratory, or MBL, and
the U.S. Bureau of Commercial Fisheries and Aquariums. The three institutions, each eminent in its own right, tend to coexist as separate little
universes. Engineering and physical sciences set the tone at WHOI, cell and

molecular biology dominate the MBL, and applied ecology is the business of Fisheries. But while the professional—and, alas, the social—spheres of the three enclaves do not overlap greatly, the scientists and technicians of Woods Hole are united by perhaps the most ancient of terrestrial diseases: sea fever. Physics, biology, or ecology can well be studied under the pines of Duke or the ivy of Princeton, but folks at Woods Hole seem drawn to the seaside by the kind of urges that moved the ancients to worship a "separate deity, and own brother to Jove."

Neptune's kingdom has drawn scholars to many harbors: Naples, Villefranche, and Bermuda come to mind as other centers where marine science has flourished. But the New England shore has a special meaning for those engaged in voyages of discovery. From the landfall of the *Mayflower* at Provincetown to the triumph of the New Bedford clippers, the path to new worlds was by way of the sea. Perhaps it is no accident that the most dazzling of our epics is the tale of a Yankee in search of a whale.

> What wonder then, that these Nantucketers, born on a beach, should take to the sea for a livelihood. They first caught crabs and quahogs in the sand; grown bolder, they waded out with nets for mackerel: more experienced, they pushed off in boats and captured cod; and at last, launching a navy of great ships on the sea, explored this watery world; put an incessant belt of navigation round it; peeped in at Behring's Straits; and in all seasons and all oceans declared everlasting war with the mightiest animated mass that has survived the flood; most monstrous and most mountainous!
>
> Melville, *Moby-Dick*

Melville may have worked in a Manhattan counting house, but the whalers of his mind left from colder seas; Ishmael took the packet for Nantucket through the waters of Woods Hole. So too for a hundred years have marine scholars plied their craft by the shores of the Cape and its islands; the search for Leviathan—as fish, or ship, or secret of the cell—does not seem entirely preposterous there.

Our small lab in the Whitman building of the MBL is only a hundred yards or so from the main dock of WHOI, so we were there in no time at all. A good-natured crowd of visitors, tourists, and locals were milling around the gates. It turned out that passes were required, but with a neighborly gesture the guard waved us by in our MBL T-shirts. The scene on the dock was Frank Capra in a nautical setting: the crowd had clearly assembled to welcome Jimmy Stewart back to his hometown. Reporters of all shapes and sizes jockeyed for post position, television cameras were mounted on scores of tripods, Coast Guard and naval officers strutted their stripes, dockhands in cutoffs and gym shirts lugged hawsers by the pier. Assembled on a kind of grandstand were bigwigs in blazers, officials in seersucker, and the gentry of Falmouth in pink linen slacks. The Marine Biological Laboratory was represented by a packet of students and faculty from courses in physiology, embryology, and neurobiology. (The biologists, if one judged them by details of facial toilette, seemed to have strayed in from a remake of *The Battleship Potemkin*.) There were also teenagers with spiked hair and bubble gum, fresh-scrubbed wives of the ship's crew, outdoor types from WHOI and Fisheries, greasy kids from the boatyard across the street, aproned kitchen staff from the four local restaurants, firemen, and town cops.

> But to omit other things (that I may be brief) after long beating at sea they fell with that land which is called Cape Cod; the which being made and certainly known to be it, they were not a little joyful.
> William Bradford, *Of Plymouth Plantation*

The ship was now upon us. The size of a minesweeper, its silhouette in the sun displayed great winch posts at the stern. As the vessel berthed broadside, general applause and happy cheers greeted the explorers. Near the top stood the head of the expedition, Robert Ballard. He stood the height of a hero, wearing a baseball hat with the sailboat logo of the Oceanographic Institution. Next to him stood a naval officer in khaki to remind us that the Navy had supported much of this research. A platoon of oceanographers leaned against the railings of a lower deck. They were outnumbered by grinning crew members and a few lab types with round spectacles. The air rang with shouts from ship to shore and back again as friends and relatives hoisted the kind of encouraging signs that one sees at road races. Tots and schoolkids were hauled aboard as the ship made fast. Flash bulbs popped and so did the corks from sudden champagne bottles. Oceanographers soon looked like winning playoff pitchers at Fenway Park. Kisses were exchanged and animals petted. More cheers and applause. The journalists gave way: there would be a press conference a little later. It took a while for the crowd to disperse, but we didn't wait for that since the champagne was still flowing. We crossed the street and reverted to the lab, where we spent the rest of the morning watching the dissociated cells of marine sponges clump together in a test tube.

Within the week, posters appeared all over Falmouth announcing that Robert Ballard would give two lectures at the Lawrence Junior High School auditorium on August 6 for the benefit of Falmouth Youth Hockey. At these talks he would show pictures of the *Titanic* expedition before they were released to the general news media. The pictures would include footage obtained by means of novel television cameras mounted on a little robot, Jason, that had poked down the grand stairwell of the liner. My wife and I bought tickets for the first of the talks to be given at four in the afternoon, and in preparation for this event, we scoured junk shops and bookstores for literature on the *Titanic* disaster. Although it cannot be said that we stumbled across unknown masterpieces of prose, the dozen or so accounts were reasonably accurate jobs of popular history. Written at various times between 1912 and 1985, they told pretty much the same story. Their overall congruity is due no doubt to their reliance on the same primary sources, chief of which were the records of two investigative commissions, American and English.

On April 15, 1912, at 11:40 P.M., while on its maiden voyage from Southampton to New York, the largest and most luxurious ocean liner of its age struck an iceberg at latitude 41°46' north, longitude 50°14' west, some 360 miles off the Grand Banks of Newfoundland. By 2:20 A.M. the next morning, the ship had sunk. Only 711 out of 2,201 souls on board escaped the shipwreck. A U.S. Senate subcommittee, headed by William Alden Smith of Michigan, began its hearing on April 19 and shortly thereafter reported its findings as follows:

> No particular person is named as being responsible, though attention is called to the fact that on the day of the disaster three distinct warnings of ice were sent to Captain Smith. . . .
>
> Ice positions, so definitely reported to the *Titanic* just preceding the accident, located ice on both sides of the lane in which she was traveling. No discussion took place among the officers, no conference was called to consider these warnings, no heed was given to them. The speed was not relaxed, the lookout was not increased.
>
> The steamship *Californian*, controlled by the same concern as the *Titanic*, was nearer the sinking steamship than the nineteen miles reported by her captain, and her officers and crew saw the distress signals of the *Titanic* and failed to respond to them in accordance with the dictates of humanity, international usage, and the requirements of law. . . .
>
> The full capacity of the *Titanic*'s life-boats was not utilized, because, while only 705 persons were saved [6 died in lifeboats] the ship's boats could have carried 1,176.

No general alarm was sounded, no whistle blown and no systematic warning was given to the endangered passengers, and it was fifteen or twenty minutes before Captain Smith ordered the *Titanic*'s wireless operator to send out a distress message.

The commissioners might have noted several other factors that contributed to the disaster. Whatever number of additional persons might have crowded into lifeboats, these had in any case room for only about half of those aboard (1,176 of 2,201). In addition, the two lookouts in the crow's nest had not been given binoculars with which to spot the iceberg, and once the berg was unavoidable, an error of navigation compounded the wreck. Although the design of the ship was such that she probably would have survived a head-on collision of almost any force, the first officer swung the liner hard-a-starboard, thereby exposing a broadside target for impact.

Seventy-five years of rehashing details of the *Titanic* disaster have not added much to this bare outline, although recent opinion has tended to lay a good share of the blame at the feet of the owners of the White Star Line. Social critics accuse J. Bruce Ismay and his financier, J. P. Morgan, of sacrificing safety for speed and prudence for luxury. In contrast, amateur steamship enthusiasts trace the ocean wreck to many individual flaws of naval conduct, culminating in negligence by the captain of the *Californian*. But if one is neither a special pleader nor a buff of shipwrecks, the story of the *Titanic* can be read as that of a unique, unlikely accident that was not part of a general pattern of nautical malfeasance. Only the sentimental can derive from the sinking ship an intimation of Western mortality: the wreck had no immediate predecessors and no similar accident happened again. Indeed, it is difficult to determine whether reforms instituted in response to the *Titanic* affair played a role in the remarkable safety record of ocean liners between the wars. Large ships that were faster and more luxurious than the *Titanic* made hundreds of trips in similar waters; the *Queen Mary*, the *United States*, the *Île de France*, and their sister ships lived out their useful lives without incident.

Nevertheless, over the years, a more or less constant set of moral lessons has been drawn from the disaster; these cautionary tales split predictably in accord with the plate tectonics of class and party. The first of these is captured in the popular image of handsome men in evening clothes awash on a tilting deck. The band plays "Autumn."

Said one survivor, speaking of the men who remained on the ship: "There they stood—Major Butt, Colonel Astor waving a farewell to his wife; Mr. Thayer, Mr. Chase, Mr. Clarence Moore, Mr. Widener, all multimillionaires, and hundreds of other men, bravely smiling at us. Never have I seen such chivalry and fortitude." . . .

But these men stood aside—one can see them!—and gave place not merely to the delicate and the refined, but to the scared Czech woman from the steerage, with her baby at her breast; the Croatian with a toddler by her side, coming through the very gate of death, and out of the mouth of Hell to the imagined Eden of America.

Logan Marshall, *The Sinking of the Titanic and Great Sea Disasters*

This lesson—the noblesse-oblige theme—includes the story of Mrs. Isidor Straus, who returned from her place in lifeboat No. 8 to her husband, the owner of Macy's. Taking her husband's hand, she told him, "We have been living together many years. Where you go so shall I." And the magnate refused to go before the other men. Harry Elkins Widener, grandson of a Philadelphia mogul, went to his death with a rare copy of Bacon's *Essays* in his pocket; Harvard owes not only its library but its swimming requirement to his memory. Benjamin Guggenheim, the smelting millionaire, went downstairs to change into his best evening dress. "Tell my wife," he told his steward, who survived, "tell her I played the game straight and to the end. No woman shall be left aboard this ship because Ben Guggenheim was a coward."

Then there was Major Butt, aide and confidant of President Taft. Mrs. Henry B. Harris reported:

When the order came to take to the boats he became as one in supreme command. You would have thought he was at a White House reception, so cool and calm was he. In one of the earlier boats fifty women, it seemed, were about to be lowered, when a man, suddenly panic-stricken, ran to the stern of it. Major Butt shot one arm out, caught him by the neck, and jerked him backward like a pillow. . . . "Sorry," said Major Butt, "but women will be attended to first or I'll break every bone in your body."

Whereas 140 of 144 (97.2 percent) women, and all the children in first class survived, only 57 of 175 (32.6 percent) male first-class passengers made shore. This example of social discipline served as a moral lesson for the gentry, who later went to the trenches in Flanders as if to a test match at Lord's.

The gallant behavior on the part of the moneyed class probably derived from the English code of the gentleman. On the *Titanic*, that code was honored to a remarkable degree. As the commander of the liner was going under with his ship, his last words were: "Be brave, boys. Be British!" One does not abandon the ship. That part of the legend—from out of the past, where forgotten things belong—keeps, indeed, coming back like a song. Twenty-six years after the wreck of the *Titanic*, Ernest Jones arrived in Vienna in the wake of the Anschluss and tried to persuade the aged Sigmund Freud to leave Hitler's Austria. Freud replied that he had to remain in the city where psychoanalysis was born. Leaving Vienna, he explained, would be like a captain leaving a sinking ship. Jones reminded him of the *Titanic*'s second-in-command, Lightoller, who was thrown into the water by a boiler blast. "I didn't leave the ship," he explained of his survival, "the ship left me!" Reassured by the code of the British gentleman, Freud took not only the lesson but also the Orient-Express to Victoria Station and freedom.

More recent students of the *Titanic* story have drawn a quite different set of lessons from the statistics and offer an analysis that one might call the *Upstairs, Downstairs* version of the disaster. Pointing out that the social classes were quartered on the ship as in Edwardian society at large, they find that steerage passengers fared less well than their upstairs shipmates: half as well, in fact! Of men in third class only 75 survived of 462 (16.2 percent); of women, 76 of 165 (46 percent); of children, 27 of 79 (34.2 percent). These statistics—literally the bottom line—yield another irony. Only 14 of 168 male passengers in second class, a mere 8.3 percent, survived. One might conclude that middle-class men adhered more closely to upstairs values than did the entrepreneurial folk on top deck.

Darker streaks of division mar the canvas. Many of the accounts of the time stirred up nativist sentiments, and the worst charges were leveled against dark, swarthy foreigners. Reporters grew indignant that "men whose names and reputations were prominent in two hemispheres were shouldered out of the way by roughly dressed Slavs and Hungarians." Rumors were commonplace—and have since been disproved—that violent battles took place in steerage: "Shouting curses in foreign languages, the immigrant men continued their pushing and tugging to climb into the boats. Shots rang out. One big fellow fell over the railing into the water. . . . One husky Italian told the writer on the pier that the way in which the men were shot was pitiable!"

Another rumor of the time is contradicted by later accounts: "An hour later, when the second wireless man came into the boxlike room to tell his companion what the situation was, he found a negro stoker creeping up behind the operator and saw him raise a knife over his head. . . . The negro intended to kill the operator in order to take his lifebelt from him. The second operator pulled out his revolver and shot the negro dead."

Those often-told dramas of the *Titanic* can be squeezed for the juice of class struggle, but the real fear of the time was not of social unrest. Led by the great populist William Jennings Bryan, the moralists found their true target: the enemy was luxury, luxury and speed. "I venture the assertion that less attention will be paid to comforts and luxuries and . . . that the mania of speed will receive a check," said Bryan.

Speed and comfort are among the declared goals of applied technology; those who worry about those goals—like Bryan—tend to worry about technology. For seventy-five years, those uneasy with machines have used the image of the *Titanic* to decorate the Puritan sampler that "pride goeth before a fall." The proud *Titanic* was 882 feet long—almost three football fields; contemporary illustrations show her as longer than the height of the Woolworth Building. She had a swimming pool, a putting area, squash courts, a Turkish bath, a Parisian café, palm-decorated verandas, a storage compartment for automobiles, and a full darkroom for amateur photo buffs. In the hold were hundreds of cases of luxury consignments, which ranged from thirty-four cases of golf clubs for A. G. Spalding to twenty-five cases of sardines and a bale of fur for Lazard Frères. Larder and beverage rooms stocked 25,000 pounds of poultry and 1,500 champagne glasses. This splendid, "unsinkable" hotel was powered by engines that could generate 55,000 horsepower. Rumor had it that she was not far from her maximum speed of 25–26 knots per hour when she hit the iceberg; other hearsay had it that Captain Smith was going for a transatlantic speed record. The pride of speed was blamed for the fall of the *Titanic*.

Journalists complained that "subways whiz through the tunnels at top speed; automobiles dash through the street at a speed of a mile in two minutes, and ocean liners tear through the water," but it was the clergymen who had their field day on the Sabbath after the disaster. Technological pride took a beating from the Reverend William Danforth of Elmhurst, Queens, who blamed "the age of mania for speed and smashing records. The one on whom one can fasten the blame is every man to whom all else palls unless he rides in the biggest ship and the fastest possible. He will be guilty in his automobile tomorrow." The pulpits of all denominations were united in teaching the Puritan lesson. They were hard on the pride of luxury, as manifest in the squash courts, the putting area, and the swimming pool. Had William Bradford himself been alive, he would have been the first to see the luxury steamer as "a right emblem, it may be, of the uncertain things of the world, that when men have toiled themselves for them, they vanish into smoke." The leader of the Ethical Culture Society, Felix Adler of New York, was alive enough to voice the sentiment, "It is pitiful to think of those golf links and swimming pools on the steamship which is now 2,000 fathoms deep." And Rabbi Joseph Silverman of Temple Emanu-el was of the same mind: "When we violate the fundamental laws of nature we must suffer."

In the decades since 1912, the *Titanic* has ranked high on the list of violators of fundamental law (applied-technology division). Fans of natural law put the story of the steamship right up there with the flights of the *Hindenburg* and of Icarus, the building of the Tower of Babel and the Maginot Line. Not long ago, the space shuttle *Challenger* joined those other violators. In our most recent mythology, *Challenger* and the *Titanic* have become linked in the popular mind. Both craft were the largest and fastest vectors of their kind, both were the darlings of general publicity, both carried the banners of Anglo-Saxon pride; both voyages went haywire for almost mundane reasons. In the hagiography of disaster, the binoculars absent on the crow's nest of the liner and the faulty O rings of the booster rocket have both been offered as examples of how the best of our science is in bondage to chance—or retribution.

On August 6, when we finally went to hear Ballard speak to the townsfolk of Falmouth on his discovery of the *Titanic*, I was sure that memories of the recent *Challenger* disaster were not far from the minds of many. That summer, with NASA grounded, the discovery of the *Titanic* 12,000 feet beneath the sea must have engaged sentiments in an American audience deeper than those of hometown curiosity. It seems unlikely that the community turned out in overflow numbers because of its concern for the traditional themes of *Titanic* literature. One doubts that the seats were packed with citizens who wished to hear replayed the moral lessons of noblesse oblige, the social notes of *Upstairs, Downstairs*, or the canons of technology's pride and fall. No, one might argue that the people of Falmouth went to hear the technical details of how a captain from Cape Cod

tracked down the largest, most elusive object beneath the waves: *Titanic,*
the Leviathan.

> For the buckling of the main beam, there was a great iron screw the
> passengers brought out of Holland, which would raise the beam into
> his place; the which being done, the carpenter and master affirmed that
> with a post put under it, set firm in the lower deck and otherwise bound,
> he would make it sufficient.
>
> Bradford, *Of Plymouth Plantation*

> The whale line is only two thirds of an inch in thickness. At first sight,
> you would not think it so strong as it really is. By experiment its one
> and fifty yarns will each suspend a weight of one hundred and twenty
> pounds; so that the whole rope will bear a strain nearly equal to three
> tons. In length, the common sperm whale-line measures something
> over two hundred fathoms.
>
> Melville, *Moby-Dick*

> The echo on our sonar indicated that we were approaching bottom, at
> a little more than 12,000 feet. Larry released one of the heavy weights
> on the side of the *Alvin*, and our descent slowed. Soon in the spray of
> lights under the submersible, I could see the ocean floor slowly coming
> closer, seeming to rise toward us, rather than our sinking to it. Pumping
> ballast in final adjustments, Larry settled us softly down on the bottom,
> more than two miles below the surface.
>
> Robert Ballard, in *Oceanus*

In the logbook style of his Yankee predecessors, Ballard here describes
an early training dive of the deep submersible craft *Alvin*. And in this same,
informative fashion, Ballard went on that summer afternoon at the junior
high school to detail his two trips to the *Titanic* site. He spoke of the
principles of oceanography, of the ground rules of hydrodynamics, and of
how optics and sonar had been used to establish the site of the wreckage.
He told us something of his decade-long career in manned submersible
craft: of continental creep and hot geysers on the Mid-Atlantic Ridge. He
told of the dark, sterile sea two miles beneath the surface and of the rare
creatures that inhabited those depths. He acknowledged his French collaborators, without whom the wreck could not have been found, and praised
the technicians of Sony who fashioned the pressure-resistant TV apparatus
of the robot Jason. And then we saw film clips of the second voyage to the
wreck of the *Titanic*, taken by Jason and its larger partner, the submersible
Argo.

By the blue lights of Argo's cameras, we saw the decks, the winches, the bridge. The stern had become undone and the huge boilers had been scattered across the ocean floor. We saw stalagmites of rust and intact bottles of wine. We went with Jason into the cavern of the great staircase and marveled at the preservation of metalwork, silverware, and leaded glass in that cold sea. We had entered the belly of the whale.

Guided by our Ahab-Ishmael we returned to the surface as the submersibles were retrieved and stowed. Ballard suggested that these pictures tended to discount the hypothesis that the iceberg had torn a great gash in the liner's side and that instead the welds had popped from the impact. The ship's hull had cracked like a nut. But his peroration was not devoted to further anecdotes of how sad it was when the great ship went down. Ballard ended with the message he had brought to the shore for a decade: the ocean and its depths are a frontier as awesome as space itself.

When the lights came on, Ballard answered questions from his fellow townspeople. Yes, he was pleased that Congress had passed a resolution that would make the wreck site a permanent monument to the victims. No, he thought that salvage was impractical; we have better examples of Edwardian artifacts and he doubted that a few chamber pots and wine bottles would be worth the gross expense of raising the ship. Yes, the *Titanic* trip was, in part, an effort to organize support for the programs of deep ocean science.

The applause that followed was long and loud. The happy crowd, from starry-eyed teenagers to oldsters with aluminum walkers, emerged into the sunlit afternoon looking as if each had been given a fine personal present. Many of us from the Woods Hole laboratories shared that sentiment; town and gown of Falmouth had been joined in a victory celebration for science and technology. The reception of *Titanic Rediviva* reminded us that science appeals to people not only for the gadgets it invents but also for the answers it provides to the most important questions we can ask: What happens when we drown? How deep is the ocean? How bad is its bottom? How fierce is the whale?

After his first voyage, Ballard had told the House of Representatives' Merchant Marine and Fisheries Committee that he was neither an archaeologist nor a treasure hunter. "I am," he told the congressmen, "a marine scientist and explorer. I am here to point out that the technological genius most Americans are so proud of has entered the deep sea in full force and placed before us a new reality."

Influenced no doubt by Ballard's publicity on television, in newspapers, and in magazines, not all of the scientists at Woods Hole shared my enthusiasm for the *Titanic* adventure. At a number of gatherings later that summer, one heard the nasty buzzing of such adjectives as "publicity-seeking," "grandstanding," "applied," "not really basic," "developmental," and, perhaps most damning, "anecdotal." It has been no secret to the public at large since *The Double Helix* that scientists are no more charitable to each other than are other professionals; novelists, investment bankers, and hairdressers come immediately to mind. But the detractors of the *Titanic* adventure were not only upset by Ballard per se. The naysayers also complained that technology rather than science was becoming imprinted on the collective unconscious of television. Some of those most vexed by Ballard's sudden prominence had themselves made major findings in the "new reality" of genetic engineering, neurobiology, and immune regulation. Were not their achievements also part of the "technological genius most Americans are proud of"? They argued that their contributions to basic science will affect the world of the future in ways more fundamental than adventures on the ocean floor.

But that reasoning strikes me as very self-serving. Historians of science and technology assure us that it is difficult to decide whether public practice follows private theory or whether the opposite is true. It is, they teach us, hard to know where technology ends and science begins. Moreover, really important discoveries, whether basic or applied, influence our social arrangements as they in turn are influenced by them. Such discoveries tend to attract attention. The Spanish court did not ignore the voyages of Columbus, nor did Galileo fail to catch the ear of the Vatican. Einstein's relativity was featured in headlines by *The New York Times*, and polio was conquered in public. When one of the new dons of DNA discovers something as spectacular as the wreck of the *Titanic*, his lectures are likely to fill auditoriums larger than that of a junior high school on the Cape. When she finds the vaccine for AIDS or solves the riddle of schizophrenia, purists will probably carp at the publicity, but I want to be in the audience to hear her grandstanding.

> And still deeper the meaning of that story of Narcissus, who because he could not grasp the tormenting, mild image he saw in the fountain, plunged into it and was drowned. But that same image, we ourselves see in all rivers and oceans. It is the image of the ungraspable phantom of life; and this is the key to it all.
>
> Melville, *Moby-Dick*

When Ishmael—or Melville—emerged from the sinking *Pequod* to tell the story of Moby-Dick, he told us as much about the science of whales as about the descent into self. Ballard's tale of the *Titanic* is not only the story of deep ocean science but also a tale of memory, of desire, and of that search for the ungraspable phantom of life that some have called Leviathan.

24

DAUMIER AND THE
DEER TICK

Daumier had rented a modest apartment in the Île St.-Louis, on the quai d'Anjou. Over the top of it extended an enormous attic which in the eyes of the landlord was quite valueless. For in this once noble but now deserted neighborhood, space at that time counted for nothing. All the artist had to do to make himself a vast and splendid studio was to have this attic plastered and open up large windows on the roof where there had been a skylight. This attic he had connected with his rooms by means of a frail and elegant spiral staircase.

THÉODORE DE BANVILLE, *Petites études: Mes souvenirs*

Nowadays, the frail and elegant staircase has long been boarded over and Daumier's old studio—still a rental property—is reached by way of the main stairwell. One night last winter, my wife and I found ourselves climbing those ancient stairs to attend a song recital by Tobé Malawista. The concert was in the nature of a farewell party given by Tobé and Stephen Malawista for their colleagues and friends at the end of a sabbatical year in Paris; the Malawistas had the good fortune to live in Daumier's studio. Steve is an old colleague of mine. A fellow rheumatologist, he is best known for having headed the group at Yale that first described Lyme disease and went on to identify its cause and cure. Last year he had been studying molecular biology at the Institut Pasteur, and Tobé had worked hard at her voice lessons.

After the dark of the wind-swept quay and the gloom of those five flights of wooden stairs, we were glad to arrive at the brightly lit studio. A chatty crowd of sixty or more faced a temporary platform that had been constructed at one end of the loft for the soprano and her accompanist. We barely had time to find two unoccupied places in the rear before friendly waves and bilingual salutes gave way to an opening round of applause. The soprano and pianist made their entrance.

The first piece was a fluttering aria by Scarlatti, "O cessate di piagarmi" ("Oh, stop tormenting me"), sung by Tobé in a far from amateur voice that sparkled in the middle registers. The audience beamed in approval as their hostess showed not only charm but command. Italian gave way to a stately air of Handel: "Oh, had I Jubal's lyre"; the extensive repeats of the Baroque master gave me time to look around the crowded studio.

Daumier's attic, as high but somewhat longer than a squash court, had been altered since Banville's visit by additions of a kitchen and bedroom. The axis of the studio was defined by two great skylights, north and south; at one end of the room a metal staircase gave access to the roof. The plaster walls were hung with several recent oils of the Fauvist persuasion. Much of the present-day furniture had been cleared to accommodate the audience, seated on rows of rented chairs. There was no trace of the square, black enameled stove that was the chief item of fixed furniture in Daumier's time and around which his visitors and assistant would sit, each with a "liter or so in hand" as the brothers Goncourt describe.

Scanning the audience assembled in their paisley, suede, silk, loden, denim, and tweed, I spotted some of Steve's équipe from the Pasteur, molecular biologists from suburban Villejuif, hematologists from New York Hospital, rheumatologists from Brooklyn; later I met lawyers, doctors, musicians, art dealers, and philosophers. We were an assembly of both sexes, several nations, and many cultures. How appropriate it all seemed: Were we not in spirit the very subjects of Daumier's art? Were we not, in Henry James' words, our century's version of the "absolute bourgeois" to whom Daumier held up the "big cracked mirror" of caricature.

Daumier moved to the Île St.-Louis in 1846 after his late marriage to a young seamstress. He was thirty-eight and prepared to settle down; he had already known both fame and misfortune as a popular caricaturist. In the 1830s Daumier's lithographs of King Louis Philippe as Gargantua or a fat pear had cost the cartoonist six months in the political prison of Ste.-Pélagie. But now, in the waning days of the July Monarchy—the regime that Tocqueville called the least cruel and most corrupt in French history—Daumier was firmly married and gainfully employed. He was commissioned by Charivari, an independent four-page daily, to produce several full-page

lithographs each week, sharing the third-page slot with Grandville's me-
nagerie and the kinder, gentler spoofs of Gavarni.

The Île St.-Louis had much to recommend it to Daumier. In the first
place, rents were cheap. The old houses on the quays, which dated from the
age of Louis XIII, were by and large in disrepair. Tenements less grand than
these were overcrowded by workers, artisans, and small shopkeepers who
had been displaced from the new constructions around the Hôtel de Ville
and the Cité. Ancient neighborhoods were toppled in the name of progress
and sanitation, while the building speculators of the July Monarchy followed the regime's advice of "*Enrichissez-vous!*" The gentrification of Paris
moved ever westward, a trend not reversed until the 1970s. The Île St.-Louis
had settled into the picturesque squalor of a fringe neighborhood.

Daumier was attracted to the Île not only for its raffish life, but also for
its spectacular views. No painter of rural scenes, Daumier loved best the
reassuring landscape of Parisian masonry. We might assume, from its
frequent appearance in his lithographs, that his favorite prospect was right
outside his front door, where the five arches of the nearby Pont Marie dip
into the Seine like the tines of a graceful fork. A corner away, on the ancient
rue Femme-sans-tête was a riverside bistro where ragpickers, bargemen,
and river navvies congregated. Daumier, a lifelong sot, often drank with
them until he was himself *sans tête*. He did not find it inconvenient that the
large wholesale wine market of Paris lay directly across the Seine to the
south. He describes lurching home, bottle-laden, along the narrow lanes of
the Île, banging his shoulders against the houses on either side. Charles
Baudelaire showed that side of him—the Femme-sans-tête, ragpicker
side—in a poem the first draft of which he gave Daumier as a present:

> On voit un chiffonnier qui viént, hochant la tête,
> Buttant, et se cognant aux murs comme un poète,
> Et sans prendre souci des mouchards, ses sujets,
> Épanche tout son coeur en glorieux projets.

> [One sees a ragpicker coming, bobbing his head,
> Stumbling and knocking against the walls like a poet,
> And, oblivious to police spies, his subjects,
> Pours out his grand designs straight from the heart.]

Although Daumier and Baudelaire might appear as two antipodes of the
French cultural globe, it seems likely that Daumier chose his studio on the
quai d'Anjou because his friend Baudelaire lived a few doors away at the

Nadar, *Honoré Daumier*. Museum of Modern Art, New York.

Hôtel Lauzun. The two differed wildly in appearance. Daumier was large, his face puffed with drink and sausage; only his eyes were said to betray wit and resolve. He was clothed and coiffed in the slightly tatty style of the absolute bourgeois; the photograph by Nadar makes him look like his contemporary Claude Bernard, the self-made professor of physiology, giant of the Collège de France. Baudelaire, pale, intense, and clean-shaven, was either dressed to the teeth in the latest linen of upper bohemia or else disheveled in rags, depending on his fortunes. Daumier's domestic arrangements were tranquil, the poet's life was awash in drugs and disease. Nor could two fictions have been further apart than the bourgeois Robert Macaire of Daumier's cartoons and the jaded *"roi d'un pays pluvieux"* of Baudelaire's poem. And what could be more remote from the daily spoofs of *Charivari* than Baudelaire's eternal vision of the poet as dandy? *"Le dandysme est le dernier éclat d'heroîsme dans la décadence . . . comme l'astre qui décline, il est superbe, sans chaleur et plein de mélancolie"* ["Dandyism is the last as of heroism amid decadence . . . like a waning star, it is superb, without warmth and full of melancholy"]. *"Le dandy doit aspirer d'être sublime sans interruption. Il doit vivre et dormir evant un miroir"* ["The dandy should try to be sublime without interruption. He should live and sleep before a mirror"].

But Daumier and the dandy were lifelong friends and drinking companions. In the early 1840s Baudelaire and his mistress, the "black Venus" Jeanne Duval, frequented Daumier's digs on the rue de l'Hirondelle and in the mid-1850s it was Baudelaire who sat with an anxious Mme. Daumier when the artist was ill. It is also Baudelaire to whom we owe the best contemporary essay on Daumier's caricature: *"Quelques caricaturistes."* The poet understood that Daumier had seized on the subtle textures made possible by the new medium of the lithograph to gain chromatic effects as rich as those of oil painting. Baudelaire pointed out that Daumier had used black and gray as colors to describe a *physiologie* of bourgeois life: the artist had found a new medium to depict a new class. Daumier's middle-class contemporaries of the 1840s gobbled up innumerable volumes devoted to human *physionomie, mimicologie,* and *analyses philosophiques et physiologiques;* the caricaturist offered a convenient guide to their own zoology. Baudelaire spoke of the Daumier bestiary: *"Feuilleter son oeuvre, et vous verrez défiler devant vos yeux, dans sa réalité fantastique et saisissante, tout ce qu'une grande ville contient de vivantes monstruosités"* ["Leaf through his work and you will find paraded before your eyes, in their striking and fantastic reality, all the living monsters that a large city contains"].

Only some of what a large city had to offer in the 1980s was present in Daumier's studio on the night of the concert. But if our sample of the twentieth-century bourgeoisie was short on living monsters, we lacked not for molecular biologists or arthritis doctors. Sons and daughters of the middle classes, we were very tame monsters indeed. Nothing more exotic was on view than the occasional mod haircut of a geneticist or the ostrich boa of an editor. Another stirring Handel aria sprang tunefully from Tobé Malawista's throat: "O sleep, why dost thou leave me?" No such objection arose from the loden-jacketed chap before me; his unlit pipe hung precariously from his lips. He resembled the youthful self-portrait of Gustave Courbet that is often called *The Man with a Pipe* (1848). The half-smiling artist of the portrait has shut his lids as if in drugged reverie: the pipe dangles from his lips. The picture is not of a tobacco smoker.

By the time Daumier had established his atelier on the quai d'Anjou, Baudelaire, Gautier, and Courbet had been introduced to an artificial paradise three doors away. The wealthy *patron* of the Hôtel Lauzun, Émile Boissard de Boisdevier, was a sometime Salon painter (1835) who had founded the Club des Haschichins. The French at the time ruled much of North Africa, and the favorite weed of the dark continent ruled the Île St.-Louis. Boissard's soirées at the Hôtel Lauzun attracted much of the avant-garde of the day, and his visitors became subjects for nightly experiments in intoxication by wine, hashish, or discourse. The laboratory notes of these experiments were kept by Baudelaire: *The Artificial Paradise* contains an almost clinical description of the isolating effects of cannabis compared to the socializing graces of alcohol. The distinction has not been better drawn since. The dandy before the mirror has watched his mind inflate, he has performed experiments at the back of his skull: *"L'homme n'échappera pas à la fatalité de son tempérament physique et moral: le haschich sera, pour les impressions et les pensées familières d'homme, un miroir grossissant, mais un pur miroir"* ["Man cannot escape the fate dictated by his physical and moral temper: hashish will [be found to] be an enlarging mirror for man's familiar thoughts and fancies, but a clear mirror"].

Daumier and Baudelaire—caricaturist and dandy, cracked mirror and enlarging glass, bourgeois and bohemian—had more in common than their biographies suggest. One need not look very far to find the dandy in Daumier: one finds it in the cadenzas of crayon that define the flutter of a lawyer's robe, in the quivering jowl of a connoisseur, the undulating cravat of a doctor. But his dandiest creation was his extravagant police spy, M. Ratapoil, stool pigeon for the emperor: one of the *"mouchards, ses sujets."* The flair of Ratapoil's moustaches, the ruined finery of his morning coat

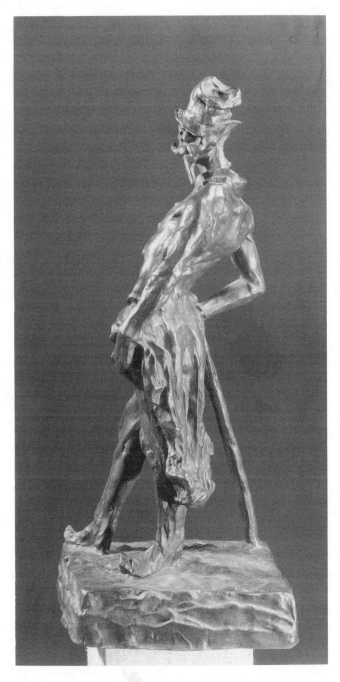

H. Daumier, *Ratapoil*. Musée d'Orsay, Paris. © Photo RMN.

and high chapeau brought out virtuoso passages of clay calligraphy. In Ratapoil, the modelling is in the service not only of character but also of self-display.

Nor is the sound of the city missing from the seductive throb of *Les Fleurs du mal*. Baudelaire appears to have set the urban stage for those dandies who followed his lead: the brothers Goncourt, Oscar Wilde, our own Tom Wolfe come to mind. In the works of the dandy, concern with silk and self shares the page with beggars and madhouses, thieves and jails, whores and hospitals, "*où toute énormité fleurit comme une fleur*" ["where every outrage blossoms like a flower"]. The dandy is, of course, the ultimate realist of city pleasure. In this he differs from the Romantic, whose major allegiance is to the countryside. Thoughts of the Romantics came to mind as the soprano recital continued. A group of Schubert songs began with the rich warble of Heine's "*Das Fischermädchen*." In the background, the piano rumbled its wavelike comment on the lyrics: The poet invites a young fishermaiden into the ebb and flow of his heart. As this *Lieder* sequence went on, I reckoned that a good many of the scientists in the room had worked with another sort of leader sequence: that run of DNA which dictates the ebb and flow of gene transcription in man and microbes.

Microbes reminded me why Romantic fishermaidens would not have favored the ebb and flow of the Seine in Daumier's day. The river was used as a commercial waterway, a public bath, a stable, a laundry—and a sewer. Parisians of the mid-century saw no connections between their use of the water and the spread of disease. Gustave Geoffroy (1878) describes the Seine beneath Daumier's window:

> *La Seine devient une paisible rivière de campagne bordée d'arbres, avec un large greve. . . . Au loin, des chalands dorment sur une eau verte. . . . Les chevaux que l'on baigne, solides bêtes de travail, émergent de l'eau en massive sculptures. Des enfants courent, jambes nues; les pécheurs à la ligne sont immobil. Des blanchisseuses pliants sous le faix remontent les escaliers de pierre. Des ouvriers et des bourgeois, coude à coude, regardent les remous et ses sillages. Tout cela sur un fond des maisons blanches, rousses et grises du quai, au toits inégaux, aux fenêtres irrégulières.*
>
> [The Seine becomes a peaceful country river with large banks bordered by trees. . . . In the distance, barges rest on green water. . . . The horses, solid beasts of labor, rise from the water after their bath like massive statues. Children run by bare-legged; the fishermen are immobile at their lines. The washerwomen, listing under their burdens, climb back up along stone steps. Workers and bourgeois, side by side, look at the

swirling water and its wake. All this against a backdrop of white, rust, and gray houses of the quay, with their uneven roofs and irregular windows.]

But the horses, children, fish, and laundry were doused by a river fed by open sewers: five miles of open effluent. I do not believe it an accident that the years of civil revolt in France (1830–32, 1848–50) coincided with peaks of cholera epidemics in the summers. The June revolution in the cholera summer of 1848 was a more desperate affair than the fraternal uprising that toppled Louis Philippe in February. By June, the workers ran not only to the barricades. The ensuing short-lived republic (1848–51) was also threatened by unrest and disease. Tocqueville pointed out that in June of 1849 "civil war, always a cruel thing to anticipate, [was] much worse combined with the horrors of the plague. For cholera was then ravaging Paris. . . . A good many members of the Constituent Assembly had already succumbed."

Many of those who could escaped from Paris. Daumier's friend Millet (*The Man with the Hoe*) fled with his family to set up his studio in the country, and it is a footnote to the period that the Barbizon School was founded in response to cholera. In June 1849 Courbet was stricken by the disease and after his recovery returned to his native Ornans, where he painted his masterpiece *The Burial at Ornans*. Again I doubt it a coincidence that his magnum opus depicts a funeral in which black pigments dominate an icon of rural death.

Indeed, it might be said that the obligatory black costume of the nineteenth-century bourgeois served as a metaphor of death by infection. Marxist historian T. J. Clark has presented a remarkable analysis of *The Burial at Ornans* in which he argues that Courbet used black costumes, social groupings, and other iconographic devices to demonstrate—among other things—that the Courbets of Ornans were not simple peasants but members of the rural bourgeoisie. One definition of a bourgeois was that unlike a worker, peasant, or bohemian he could afford to pay for his own funeral, and Baudelaire himself pointed out that the black daydress of the bourgeois made him look as if he were perpetually attending someone's funeral. Before the 1830 outbreak of cholera, which spread from Provence to the capital, the *habit noir* was uncommon; by the 1848 epidemic it had become battle dress on the boulevards. Baudelaire's bourgeois was busy attending the funerals of cholera victims.

G. Courbet, *The Burial at Ornans*. Musée d'Orsay, Paris. © Photo RMN.

During the February revolt Baudelaire urged Republicans to the barri-
cades; gun in hand, he called for the death of General Aupick, hero of
Waterloo and head of the elite École Polytechnique. The gesture, worthy of
Mark Rudd or Oedipus, was not often repeated after the cholera days of
June. By summer's end, Baudelaire, who happened to be the stepson of
General Aupick, had left for the clean air of the country. Only Daumier
remained in Paris throughout the cholera summers. His print of cholera—in
the black and white of an etching, not the grays of lithography—shows
deserted streets with an emaciated dog, a pair of stretcher-bearers depart-
ing, a victim in black dying in the gutter, a passing hearse with two in train.

The poor fought the new bourgeois republic as they fought disease, and
they equated the two. Clark has translated the complaint of a M. Dupont,
ally and friend of Baudelaire:

> La Réaction sur nos murs
> Étale son aile livide;
> Chassons les miasmes impurs
> Dégageons le marais putride.
> Nous nous plaignons du choléra
> Il nous guette dans notre gîte:
> Un coup de vent le balaira
> Hurrah! les morts vont vite!

> [Reaction spreads its livid wings
> Over our walls;
> Let's chase away the dirty specters,
> Drag ourselves from putrid swamps.
> We may complain about cholera
> It's waiting in our homes:
> One whiff of fresh air will clear it away.
> Hurray! The dead go quickly.]

Predictably, the middle classes took a different view of cholera. A popular
engraving of the time entitled *Two Scourges of the Nineteenth Century, Socialism and Cholera* shows a wraith reading Proudhon's socialist newspaper,
while at the foot of the guillotine cholera plays a flute carved of human
bone. A friend of Daumier, Vetex, was commissioned by the Republic to cast
a statue when the epidemic had waned; nowadays it sits in an alcove of the
Salpêtrière. It shows an overwrought—and underdressed—figure of Paris
thanking God for saving the city from cholera. Intervention by the deity

was not very effective: 75,000 Parisians died, the bulk of those in the poorer quarters of the town.

Haussmann's urban renewal during the Second Empire included reconstruction of the city's sewer system. Thus when cholera next threatened the capital from Baltic and Mediterranean ports, Paris was spared. Not only improved sanitation but also strict quarantine measures saved the city. The disease was not, however, on its way to elimination until Robert Koch isolated the causative microbe (a vibrio) in 1884 and rediscovered its waterborne route of transmission. Footnotes to literary history tell us that Émile Zola's father, a civil engineer, died before he could finish building a clean water system for the city of Avignon in the wake of the 1830 epidemic, and that Marcel Proust's father, a public health doctor, played a major role in the sanitary reforms that kept cholera from French borders. We might conclude that literature is the child of hygiene.

Clean water and literature, if not hygiene, were on our minds as Tobé swung into the last song before intermission. Appropriately enough, it was Poulenc's *"Air champêtre"* set to a fey ode by Jean Moréas. The symbolist poem addresses itself to a gurgling country stream:

> O nymphe à ton culte attache
> Pour se mêler encore au souffle qui t'effleure
> Et répondre à ton flot caché . . .
>
> [O nymph, I would join your cult,
> To mingle again with the breeze that caresses you
> And to respond to your hidden waters . . .]

There are no nymphs in Baudelaire, no running brooks in Daumier. Between the Romantic ebb of Heine and the symbolic flow of Moréas, a generation of Parisian realists and dandies defined the space of urban art, a culture of the new based on observation. In Daumier's art, the observations were most acute when his medium was that of graphic journalism: its fluent sketches, impromptu poses, multiple copies. His paintings, on the other hand, were patterned after an older vernacular of easel painting. He had looked long and hard at the Old Masters in the Louvre and learned his lessons from the savviest printmaker of them all, Rembrandt. The chiaroscuro of Rembrandt's canvases agreed with the Dutch passion for etching; Daumier based his smallish oil paintings on the journalistic conventions of lithography: the closeup grimace, the backlit figure, the fuzzy corners. Daumier, the caricaturist, never quite got the hang of all that grand painting, of how to fill each meter of canvas in the conventional manner of the Salon painter. Delacroix accused him of having "trouble finishing," and indeed, Daumier never quite completed the few major pieces commissioned from him by the Republic. In most of Daumier's oils, the process of painting remains undisguised; the image remains almost unfinished. Black, lithographic lines trail off in crayonlike fashion over scrumbled washes of ocher, umber, and buff.

Daumier's experiments in oil seem to work best when their subject matter matches that of his daily work for *Charivari*: the people of Paris. For me the most perfect of Daumier's paintings is the little *Washerwoman* in the Musée d'Orsay. Immediately to the left of the tricky entrance to the museum is a room devoted to Daumier; his small paintings and clay figurines are eclipsed by the giant spaces of the former train terminal. But the washerwoman manages to hold her own against all the arches and statuary. She is brushed in dense lithographic strokes and seems as solid as the masonry of the quai d'Anjou on which she treads. She and her child are shown climbing up to the street from the riverbank below. As in a print, the figures are posed in silhouette against the creamy houses of the right bank; they are framed by the stone walls of the quay. Indeed, since the oil is almost exactly the size of a lithographer's stone (49 × 33 cm), mother and child may be said to have stepped from the artist's stone to the pavement of the street. We do not see the river: we infer its presence from the geometry of the site. The composition may derive from Baudelaire's comment to Daumier after a day in the country: "*L'eau en liberté m'est insupportable; je la veux prisonnière aux carcans, dans les murs géométriques d'un quai*" ["I find free-flowing water insufferable; I want it imprisoned, yoked by the geometric walls of a quay"].

Since the site described by both painter and poet was the quay right outside the studio where the concert was held, I resolved to look at it during the short intermission. Ample halftime applause now rewarded Tobé and Mme. Aïtoff, her accompanist. There was much moving about as several rounds of amiable greetings and general introductions were exchanged. We were assured that a first-rate collation would be served at the end of the concert; there wafted from the kitchen scents that would have troubled the *physiologie* of M. Brillat-Savarin himself. I threaded my way through the bustle to a window overlooking the quay d'Anjou. The streets were well lit and the Seine sparkled from a hundred lights: a few cars scuttled along the motorway across the river. The Pont Marie was there on the left—just as in Daumier's lithographs—and here, just in front of the studio, was the little break in the quayside wall that marked the beginning of the stone staircase

H. Daumier, *The Washerwoman*. Musée d'Orsay, Paris. © Photo RMN.

down to the river. Here the washerwoman and her child had climbed up
and down, dawn and dusk. Here Daumier had caught them once and for
all. Here Baudelaire had seen the same scene, and he wrote verses as free
of sentiment as the streets of Paris:

> L'Aurore grelottante en robe rose et vert
> S'avançait lentement sur la Seine déserte.
> Et le sombre Paris en se frottant les yeux
> Empoignant ses outils, veillard laborieux.

> [Dawn, quivering in robes of rose and green,
> Crawls slowly over the empty Seine.
> And Paris, rubbing her eyes in the dark,
> Picks up her gear like an old workwoman.]

> "Le Crépuscule du matin," 1857

It struck me that both Daumier and Baudelaire were not simply observing
the Paris scene in Washerwoman or "Le Crépuscule du matin." Their observations had been modified not only by the conventions of art but by the social
concerns of the new realism. Realism in its urban phase was a meliorist
experiment in the sense that city life itself was an experiment in nineteenth-
century Paris, or in any other city for that matter. Across the river, in his
Introduction to the Study of Experimental Medicine, their contemporary Claude
Bernard had proclaimed that: "L'observation est l'investigation d'un
phénomène naturel, et l'expérience est l'investigation d'un phénomène modifié par
l'investigateur" ["Observation is the investigation of a natural phenomenon,
and an experiment is the investigation of a phenomenon modified by the
investigator"].

Looking from Daumier's studio over the dark river to the gleaming
motorway, seeing trim new buildings squeezed between floodlit landmarks
of the old Marais, watching the lights of a bateau-mouche play on ancient
façades of the Île St.-Louis, hearing strains of Charles Trenet from a transis-
tor by the riverside, I had the impression that the city experiment had
worked. Experiments of French art and science had not been the only
successful efforts of the last century. That evening, Paris itself might be said
to have been a successful experiment, an experiment carried out not only
by the likes of Bernard or Daumier but by the French people whose sense
of order sometimes coincides with the goals of equality and fraternity. The
city since Daumier seems to have survived cholera and revolution, war and
famine, occupation and betrayal, structuralism and nouvelle cuisine. I

hoped that one day the unfinished experiment of life in urban America might conclude with as much success as the Paris of that evening. An ocean away from the Manhattan of crack and crime, I was less certain than usual that modern city life was a fatal disease that required empathy rather than intervention on the part of the doctor. I was buoyed by this optimism as my host approached.

As we looked over the river, the view of water prompted me to remind Steve that my wife and I expected the Malawistas to visit our house at Woods Hole next summer, and that we would all be having dinner together within a month by the East River in New York. Steve was going to lecture on Lyme disease in a postgraduate course that I conduct each year; he had first given us a talk on Lyme arthritis in 1977. He told me he would be presenting a new aspect of the disease that should be of interest to me. A Harvard entomologist, Andrew Spielman, claimed that Lyme disease had originated on Naushon Island in Massachusetts. Since Naushon is separated from Woods Hole only by a narrow channel that joins Vineyard Sound to Buzzards Bay, and since I remembered that the tick that harbored the spirochete of Lyme disease was carried by deer, I asked Steve how the disease would have arrived on the mainland.

"Well, I guess the deer must have swum across the Hole."

Further conversation was interrupted by the flicking of fights, which signaled the end of the entr'acte. We resumed our seats as Tobé and her pianist were received with another friendly round of applause. The next two songs were by Benjamin Britten and were chirped in the staccato of folk tunes: "Oliver Cromwell is dead!" we were told, and bell-like chords echoed, "Haw-haw." Next came a song from Somerset, "O Waly, Waly." Its lyrics continued the aqueous motifs of earlier songs:

> The water is wide, I cannot get o'er,
> And neither have I wings to fly.
> Give me a boat that will carry two,
> And both shall row, my love and I. . . .
> A ship there is, and she sails the sea,
> She's loaded deep as deep can be.
> But not so deep as the love I'm in:
> I know not if I sink or swim.

Thinking about Steve's remark, I wondered whether the Naushon deer sank or swam. I had only recently learned that deer could swim at all. In fact, Prosser and Deedee Gifford, who live near us at Woods Hole, had assured us that they had seen deer swimming across the channel between Woods Hole and Naushon Island. Gothamite that I am, I was in no position to doubt the proposition that deer can swim—or fly, for that matter. Having been brought up as city-bound as Daumier or Baudelaire, I'm always ready to believe any countryman's tale of sea and forest; I'm a sucker for the tall stories in *Smithsonian* or *National Geographic* with all their wisdom of the wild. My thoughts wandered to the only deer Paris had shown me: the deer of Courbet's canvases on the level above the Daumier room at the museum across the river. Seated by the gleaming Seine that evening, with Britten's "O Waly, Waly" in my ears, I fancied that small herds of white-tailed deer might—well, why not?—have dipped their gawky limbs into the waters off the Île St.-Louis. As dawn quivered in rose and green, they had made their way downstream to the Musée d'Orsay. They had emerged, those graceful creatures, to shake the Seine off their antlers and stepped ashore to settle on the Rive Gauche. The distance they would have had to travel and the strength of current they would have encountered are roughly similar to those between Naushon Island and the Woods Hole mainland.

Deer, of course, not only can but *do* swim. And while it is not certain that deer carried the tick that carries Lyme disease from Naushon to the mainland it *is* certain that where there are no deer there is no Lyme disease. We now know that the tick (*Ixodes dammini*) spends its winters embedded in the deep fur of the white-tailed *Odocoileus virginianus*. Bambi, in Latin. The deer tend not to get the disease but they harbor all stages of the *I. dammini* tick, and therefore, in the words of Andrew Spielman, "their presence is prerequisite to the maintenance of infections transmitted by this vector."

Naushon and some of its sister islands off the eastern mainland served as wooded, brush-filled sanctuaries for deer in the three centuries during which the creatures fell prey to Indians and settlers. On the mainland, trees and deer disappeared as red man and white used them for fuel, food, and shelter. Importation of iron tools and iron weapons played a major role in leveling the land and destroying its wildlife. The scribe John Brereton, who accompanied Bartholomew Gosnold on his voyage of discovery in New England in 1602, described the landscape of the offshore islands as filled with "an incredible store of Vines, as well in the woodie part of the island;

where they run upon every tree, as on the outward parts, that we could not go for treading upon them. . . . Here also in this island, great store of Deere, which we saw and other beasts as appeared by their tracks . . ." But by 1640 deer were already on their way to destruction and, again according to Andrew Spielman, the last deer on Nantucket Island was eaten in the middle of the nineteenth century. Pictures of Woods Hole of the last century show the low coastline scalped of trees and bare of wildlife.

Only Naushon remained a deer preserve, because of the stewardship of its private owners: Presidents Ulysses S. Grant and Theodore Roosevelt visited the island and shot deer for sport. Wealthy Americans tend to favor trees on their estates; deer parks are the proper preserve of the mighty. Deer did not repopulate the mainland until the 1960s; by then, large areas of Cape Cod and the Massachusetts shore had become reforested as a result of general prosperity and the advent of the summer home. Properly zoned, much of the area had become filled with the kind of dense underbrush that resembled the landscape of the offshore islands before man and iron had changed the New World. It is therefore not surprising that deer and the deer tick first spread disease in affluent communities of the Northeast. From Ipswich in Massachusetts to Lyme in Connecticut to Shelter Island in New York, a reforested landscape had been permitted to assume its original parklike aspect. The victims of Lyme disease have been winter residents and the folk of summer who dwell in spacious homes among the trees, the shrubs, the lawns. In our epidemic, editors and agents, artists and models, writers and dandies have become victims. Cholera may have spread along the Left Bank by the water of the Seine; Lyme disease has flourished along the seacoast of upper bohemia.

Two songs by Fauré with texts by Sully Prudhomme were next on Tobé's program. More water! Repetitions of *"L'Eau murmure, l'eau murmure"* in *"Aux bords de l'eau"* were followed by dreamlike ruminations which seemed to promise release from the troubles of the world. *"Ici-bas"* also came out in favor of flowers and the sun. The music was as charming as the lyrics were banal; Tobé's warble rolled over Fauré like a harp. After a forgettable bauble of Stravinsky's, she launched into a berceuse by Tchaikovsky. The lullaby is set to a text by Maikov in which the wind, obedient to the poet's fancy, responds to his good mother's question:

"... *Qu'as-tu fait pendant*
Ces trois jours? As-tu balayé les étoiles?
As-tu repoussé les vagues?"
"Non, Maman, j'ai bercé un petit enfant."
  "Dors, mon petit."

["... What have you been doing for
Three days? Have you blown away the stars?
Have you pushed back the waves?"
"No, Mother, I have rocked a small child to sleep."
  "Sleep, little one."]

Lyme disease was discovered in the United States thanks to some good mothers in Connecticut. Mrs. Judy Mensch, who lived in a large house surrounded by woods, first encountered what was to be called Lyme arthritis in November 1975. Her eight-year-old daughter, Anne, came down with a swollen knee, which her doctor treated as "osteomyelitis." But when the symptoms persisted and the diagnosis was changed to "juvenile rheumatoid arthritis," the good mother began to suspect something wrong. Three other children on the same street had gotten severe arthritis within a year and a half; the girl next door needed a wheelchair after her third attack, and a boy around the corner and another little girl down the street were also afflicted. Mrs. Mensch began to call her neighbors and soon came up with a dozen other victims. She telephoned the Connecticut State Department of Health and spoke with Dr. David Snydman, who had just come to the state from the USPHS Centers for Disease Control (CDC) in Atlanta.

Dr. Snydman had also received a call from a Mrs. Polly Murray, whose husband and two teenage sons had come down with arthritis sufficient to put them all on crutches at the same time. Mrs. Murray lived a few miles north of Mrs. Mensch; the house was surrounded by woods. Snydman drove to Lyme to investigate, called up doctors, checked hospital records, and obtained lists of other victims, who by now seemed to exceed a dozen or so. He came up against a puzzle. Many of the cases, especially in children, seemed to fit into the clinical picture of juvenile rheumatoid arthritis, a rare and disabling chronic disease which, paradoxically, sometimes afflicts adults, but which differs from the more common, adult form of rheumatoid arthritis in not tagging its victims with a positive test for rheumatoid factor in the blood. Juvenile rheumatoid arthritis is therefore considered one of the rarer "seronegative" diseases of joints, and finding a dozen or so cases clustered in space and time was unheard of. Or was it?

Dr. Snydman put in a call to his friend Allen C. Steere, who had also just come from the CDC to Connecticut. Steere was a first-year fellow in rheumatology at Yale, and Steve Malawista was his chief. Steere's training had been in epidemiology, and Malawista urged him to get some experience at the laboratory bench. The project Malawista had set Steere when he arrived as a novice in July 1975 was to study the roles of cyclic nucleotides and microtubules in the inflammatory properties of human white cells; Steve's lab and mine have been friendly competitors in this sort of research for decades. Allen Steere was just getting started on white cells when fate beckoned.

Steere and Malawista agreed that there was something going on at Lyme that warranted dropping the white cell project. They presented their proposal to the Yale Human Investigation Committee, and with Snydman's help brought in all the known or suspected cases and performed every conceivable clinical and laboratory test. They studied the case histories and epidemiological evidence; they scoured the records and local street maps. Their original study, presented for the first time at a meeting of the American Rheumatism Association in Chicago on June 10, offered a new name: "We think that the Lyme Arthritis is a new clinical entity," the abstract concluded. By July 1976, a front-page story in *The New York Times* broke the news to the public. Written by a crack science writer, Boyce Rensberger, it gave a complete account of the ongoing study at Yale and was illustrated with a photograph of young Steere "checking a patient's knee for 'Lyme Arthritis' at Yale University School of Medicine as Dr. Stephen E. Malawista, rheumatology chief, observes." "Lyme arthritis" was in the lexicon for good.

The full manuscript of Steere and Malawista's study was published in *Arthritis and Rheumatism* as "Lyme Arthritis," subtitled: "An Epidemic of Oligoarticular Arthritis in Children and Adults in Three Contiguous Connecticut Communities"; Steere, Malawista, and Snydman were at the head of the list of authors. I have been told that Allen had originally proposed the subtitle as the full title of the article, but that Steve with his didactic flair insisted on the eponym. Be that as it may, the paper is a classic of clinical science. Differentiating the disease clearly from other known types of arthritis, the authors pointed out that "its identification has been possible because of tight geographic clustering in some areas, and because of a characteristic preceding skin lesion in some patients." The skin lesion, present in 25 percent of the fifty-one patients, was correctly identified as *erythema chronicum migrans*, an expanding red nubbin of a rash that had been identified at the turn of the century in Sweden. The skin lesion had not previously been associated with arthritis, but as Steere, Malawista, *et al.* pointed out, fever, malaise, headache, and neurological symptoms were all part of the syndrome. The Yale group correctly deduced that "although the prevalence of *erythema chronicum migrans* in the three communities reported here is not known, it is doubtful that it is so common that one quarter of the patients with arthritis would have the skin lesion just weeks before the arthritis by chance alone. Thus the authors think that both symptoms may be manifestations of the same illness." They also correctly deduced from the clustering of the disease in more sparsely settled, heavily wooded areas rather than in town centers—and the absence of a common source of transmission such as water—that "the epidemiology fits best with an illness transmitted by an arthropod vector." They had reasoned from the first that a tick might have transmitted the disease.

Whatever else the investigation led to, whatever else developed, to my mind this first description of a new disease, this exercise of reason to produce a new bit of knowledge, is an example of what we mean when we speak of imagination in science. It has given us the sensation of the new. As one rereads the article, one can hear echoes of Baudelaire's hymn to the imagination:

> *Mystérieuse faculté que cette reine des facultés. Elle touche à les autres; elle les excite, elle les envoie au combat. Elle leur ressemble quelquefois au point de se confrondre avec elles, et cependant elle est toujours bien elle-même, et les hommes qu'elle n'agite pas sont facilement reconnaissables à je ne sais quelle malédiction qui dessèche leurs productions comme le figuier de l'Évangile. Elle est l'analyse, elle est la synthése. . . . Elle a créé, au commencement du monde, l'analogie et la métaphore. Elle décompose toute la création, et avec les matériaux amassés et disposés suivant le règles dont on ne peut trouver l'origine que dans le plus profond de l'âme, elle crée un monde nouveau, elle produit la sensation du neuf.*

> [How mysterious is imagination, that Queen of the Faculties! It touches all the others; it rouses them and sends them into combat. At times it resembles them to the point of confusion, and yet it is always itself, and those men who are not quickened by it are easily recognizable by some strange curse which withers their productions like the fig tree in the Gospel. It is both analysis and synthesis. . . . In the beginning of the world it created analogy and metaphor. It decomposes all creation, and with the raw materials accumulated and disposed in accordance with rules whose origins one cannot find save in the furthest depth of the soul, it creates a new world, it produces the sensation of the new.]

> *The Salon of 1859*

Imagination in science is the faculty that permits the investigator to reassemble those snippets of fact—the spreading rash, the swollen knee, the wooded lot—according to rules that are by no means hidden in the furthest depth of the soul. We tend to use that part of our nature to give us not the rules but the courage to break them.

Within the next few years, Steere and Malawista were joined by two other young stars of Yale medicine, John Hardin and Joseph Kraft, and went on to study the disease prospectively. This time they knew where and when to look: in the Lyme area, and in the "high" season of the malady, in summer. On the basis of 314 new cases, they defined the clinical spectrum of what by 1979 was coming to be called Lyme *disease* rather than Lyme *arthritis*. The illness turned out to be more serious than originally described in 1977, but the first full description, published a decade ago, has not required significant change. Stage one of the disease is the rash, *erythema chronicum migrans*, accompanied at times by symptoms of "summer flu" with headache and fever. These symptoms and the rash itself usually go away. Stage two follows within several weeks to months. About 15 percent of patients develop neurologic problems: strange transient paralyses, confusion, and such serious consequences as Bell's palsy or meningitis. In 8 percent, the second stage of the illness involves disorders of the heartbeat; in some cases the whole heart is inflamed (myocarditis) and deaths have been reported. Arthritis—stage three—follows in 60 percent. Developing as early as a few weeks to as late as two years after the rash, the joint disease can lead to destruction of cartilage and bone. The Yale group identified markers in the serum (cryoglobulins, IgM antibodies) that predicted which patients would go on to complications and, more important, discovered that prompt antibiotic treatment at the stage of the rash would stop the disease in its tracks. Long courses of intravenous antibiotics were effective at the later stages.

Having described the disease and its treatment, and having presented good epidemiological evidence that some sort of tick living on the back of some sort of animal was the vector of some sort of microbe, the Yale group by 1979 was left with the questions of *which* specific tick, feeding on *which* animal "reservoir," carried *which* microbe that caused Lyme disease. By 1982 the answers were in. The tick, a new species of *Ixodes dammini*, was named by Andrew Spielman; the main animal reservoirs—also uncovered by Spielman and his coworkers—were the white-footed mouse and white-tailed deer; and the microbe was a spirochete that resembled the causative agent of syphilis. The spirochete, first recovered from the mid-gut of ticks, was named *Borrelia burgdorferi*, after Willy Burgdorfer of the Rocky Mountain Laboratories of the USPHS. Cats and dogs may carry the tick, birds may spread their nymphs and larvae, but if the mice or deer are killed off, there is no disease. Each of these findings required another exercise of the scientific imagination, another "observation modified by the investigator"; each produced its own sensation of the new.

Tobé had now launched into the last two songs of her recital; both the *"Coucou"* and the final *"Oublier si vite"* were by Tchaikovsky. The first of these, programmatic to a fault, fulfilled our expectations of an ode dedicated to a mistake of the avian class. The last song was lush and sentimental, and Tobé sang it in waves of lyric delight. The poem by Apukhtin mourns: *"Oublier si vite tout le bonheur vécu dans une vie"* ["To forget too soon all the happiness of one's life"].

One forgets too soon in the course of a life in science those happy, rare moments when the imagination takes over and one makes a discovery. Too often they are buried in an avalanche of disputes, of claims and counter-claims. Who had the idea first? Who did the work? Was it luck? Was it really new? Too soon one realizes that one has simply rediscovered an old observation in a new setting. Some of this happened with Lyme disease. It turns out that in 1970, a Wisconsin dermatologist named R. J. Scrimenti had reported the coincidence of *erythema chronicum migrans* with neurological symptoms in a fifty-seven-year-old doctor. Persuaded by Scandinavian reports that the rash was carried by an *Ixodes* tick that might transmit a spirochete, Scrimenti looked for a spirochete in his patient and correctly treated him with penicillin (*Archives of Dermatology*, 102 [1970], 104–5). Indeed, E. Hollstrom had already pointed out in 1958 that *erythema chronicum migrans* would respond to penicillin.

The unique spirochete was first identified by Burgdorfer after a painstaking search for a microbe in the mid-guts of ticks harvested at Shelter Island. Much of the field work in the course of the Long Island epidemic was done by a group of practitioners, pathologists, and public health doctors from New York State and from SUNY, Stony Brook. These doctors, especially Jorge Benach, Bernard Berger, and Edgar Grunwaldt, have been amply celebrated by Berton Roueché in his *New Yorker* reminiscence of the tick disease.

But the Yale group put it all together first, and got it almost all right at the very beginning. They also paved the way for a remarkable analysis of how Lyme disease is an accident of ecology and entomology. This has been brought to our attention by Andrew Spielman, who began his work with ticks in a search for the vector of an entirely different disease.

In 1969, an elderly woman who lived on a moor on Nantucket Island experienced a malarialike infection due to a tick-borne parasite called *Babesia microti*. This rare case was followed by a second one, in an acquaintance of the old woman, who had been spending the summer in a nearby Nantucket village. After six more cases of babesiosis were identified on Nantucket in 1975, other cases popped up on Shelter Island. As had the Yale group with Lyme disease, Spielman at Harvard decided that "the geographical clustering of human infections helped establish the venue for epidemiological studies."

By 1976, Spielman's group had found that white-footed mice were the major reservoir for *B. microti* and the search was on for an insect vector. A tick was suspected, but the *Ixodes* ticks found on the mice seemed to belong to a species (*I. muris*) that previously had never attached to humans. A splendid job of systematic taxonomy led Spielman and his group to conclude that the *Ixodes* tick species "collected from Naushon Island as well as most other sites located along the New England coast and on eastern Long Island proved to be a new species, later designated *I. dammini*." (Gustav Dammin was a coworker of Spielman's.) When Lyme disease and babesiosis were linked epidemiologically by their clustering in time and space, and later by their coincidence in the same patients, their shared mode of transmission via the deer tick (which harbored both microbes) became clear. Burgdorfer's identification of the spirochete in *I. dammini* followed in 1982, and soon after, there appeared in *The New England Journal of Medicine* an article entitled "The Spirochetal Etiology of Lyme Disease." Allen Steere was the first author and Steve Malawista the last; Willy Burgdorfer was among the others. The spirochete had been found in the blood, skin lesions, and spinal fluid of three patients with Lyme disease and in twenty-one ticks found in areas of Connecticut endemic for Lyme disease. It had taken from the summer of 1975 to the spring of 1983 to describe a new disease, to find out what caused it, and to learn how to treat it. Steere et al. pointed out the implications: since Lyme disease and syphilis were both caused by a spirochete, and since syphilis was the original "great imitator" of other diseases, the later manifestations of Lyme disease may mimic other conditions (heart attacks, nerve palsies) while live organisms take up residence in the spleen. In this paper, which represented the apex of their science, they did not remind their audience that Baudelaire reached the heights of his art in "Spleen."

Can we prevent Lyme disease? Spielman thinks that the best way to go about it is to get at the mice. But—sad to say—one can do this job just as well by killing off all the deer. At the request of the residents of Great Island off Cape Cod, all but one or two of the thirty-five deer that inhabited the island were shot during a two-year span. In the third year after deer were eliminated from Great Island, "the human population of that site largely became protected against infection. Whereas the 200-odd summer residents of this island previously had suffered four to eight Lyme disease infections annually, none were reported thereafter." Spielman doubts that these measures would work "when practiced in a portion of a continuous land mass." But I wonder what we will do about the deer if the epidemic of Lyme disease keeps mounting, like AIDS?

Sitting in the urban nest of Daumier's studio as Tobé came to the end of her recital I thought back on the elaborate machinery of nature and nurture that makes a microbial disease possible. For *B. burgdorferi* to make a living as a pathogen in Lyme, Connecticut, Gosnold had to have charted the Cape and Naushon Island; settlers and Indians had to have cleared the Northeast of forests and deer; Naushon had to have served as a private deer sanctuary; deer had to have swum or roamed back to the mainland of New England; mice and ticks had to have found themselves in a landscape with deer; and the economy of the United States had to have flourished enough to permit its middle classes the luxury of winter and summer homes in parklike settings. Those geographic considerations reminded me that diseases—like Daumier's art—are complicated products of civilization. Not only artists dictate the style of their time: not only microbes dictate who gets sick and from what. The canon of our new molecular biology agrees with that of the older literature: it takes a lot of global luck and human collusion for a bacterium or virus to become an agent of disease.

The intricate tricks of biochemistry required of a microbe to set up shop as a pathogen guarantee that not each microbe will have an equal chance of hurting every human. In that sense, if not with respect to its fury, Lyme disease may be said to be the AIDS of the affluent. Nowadays, while the

sad rules of the human immunodeficiency virus have been explained to
alert members of the gay community, the HIV epidemic in America affects
chiefly black and Hispanic abusers of intravenous drugs. Those folks will
never meet a deer or be bitten by *Ixodes dammini*. Nor is it likely that the
residents of Lyme, Connecticut, will have to peddle crack or heroin for a
livelihood. But although we know enough about the epidemiology of AIDS
and Lyme disease to limit their relentless spread, few of those at high risk
for either disease seem ready to change their social habits in the near future.
The slum dweller is not likely to abandon the drug culture simply on
doctor's orders; the summer resident of Cape Cod or Shelter Island is
unlikely to give up his pet cat or dog. Certainly, as Steve Malawista has
pointed out, no one will sign up to kill Bambi. We should probably not feel
superior to the Parisians bathing in the polluted Seine. When Haussmann
hacked out the sewers he did so over great opposition. Cholera was de-
feated because the fears it aroused were greater than the "rules" of social
arrangements then current. When we become afraid enough of Lyme
disease, we may be moved to greater action. I, for one, worry about the
later—as yet unknown—consequences of the disease for those now treated
simply with a fast course of antibiotics. We have a lot to learn.

As the rhythmic applause ended the concert in Daumier's studio, and
our comfortable crew of upper bohemians headed for a splendid feast, the
skylights and windows reflected our party finery. Through the glass of the
window, we must have looked like the bourgeois doctors of Daumier's
caricatures. So be it, I thought. Dominant over Western culture for a century
and a half, the bourgeois, meliorist movement we represented seems to
have gotten *some* things right. We may not have built the City of God, but
we have gotten good at treating sewage. Our social arrangements may not
be in better array than those of the Second Empire, but we can drink water
from a tap. Our painting and poetry may not be more sublime than those
of the Second Republic, but we can now kill Baudelaire's spirochete and
cure a blind Daumier. Cholera is gone, and we are busy clearing up our new
epidemics. Thanks to scientists from the Institut Pasteur we have pinned
down the virus of AIDS, and thanks to our host, Steve Malawista, and to
Allen Steere, we can identify and cure Lyme disease. Thanks to the Burgdor-
fers and Spielmans we have learned how woven into the fabric of natural
history is the health of a people.

I'm persuaded that work of this sort in the sciences requires leaps of
the imagination no less thrilling than Daumier's prints or Baudelaire's

verse. For way down there, in the furthest depth of the soul, the imagination lurks in artist and scientist alike, itching for its chance to astonish us all.

"It decomposes all creation, and with the raw materials accumulated and disposed in accordance with rules whose origins one cannot find save in the furthest depth of the soul, it creates a new world, it produces the sensation of the new. Since it has created the world (so much can be said, I think, even in a religious sense), it is proper that it should govern it." Would that it were so, *cher maître*, would that it were so.

SOURCES

Note to the reader: The date of composition for each essay is given in parentheses; dates following backslashes denote dates of major revision.

DARWIN'S AUDUBON (1998)

Alexander, P. *Commonwealth of Wings: An Ornithological Biography.* Hanover, NH: New England Universities Press, 1991.

Audubon, J. J. (engr. R. Havell). *The Birds of America.* London: Author, 1827–38.

Audubon, J. J., with W. MacGillivray. *Ornithological Biography,* 5 vols. Edinburgh: A. Black, 1831–49.

Audubon, M. R., ed. *Audubon and His Journals.* New York: Dover, 1994 (orig. 1897).

Darwin, C. *On the Origin of Species by Means of Natural Selection or the Preservation of Favoured Races in the Struggle for Life.* New York: The Heritage Press, 1963 (reprint of 6th edition, 1872).

———. *The Variation of Animals and Plants Under Domestication.* London: John Murray, 1868.

———. *The Descent of Man and Selection in Relation to Sex.* New York: D. Appleton, 1871.

———. *Autobiography and Letters.* New York: D. Appleton, 1893.

Darwin, E. *Zoonomia.* Philadelphia: Edward Eagle, 1818 (orig. 1802).

Darwin, F. *The Life and Letters of Charles Darwin.* New York: Basic Books, 1959 (orig. 1887).

Desmond, A. and J. Moore. *Darwin: The Life of a Tormented Evolutionist.* New York: Warner, 1991.

Ford, A., ed. *The 1826 Journal of John James Audubon.* New York: Abbeville Press, 1987.

Ford, A. *John James Audubon.* New York: Abbeville Press, 1988.

Fries, W. *The Double Elephant Folio: The Story of Audubon's Birds of America.* Chicago: American Library Association, 1973.

Herrick, F. H. *Audubon the Naturalist.* New York: Appleton–Century, 1938.

Lamarck, J. B. M. de. *Système des Animaux sans Vertébres.* Paris: Deterville, 1801.

———. *Philosophie Zoologique*, Vol. 1. Paris: Dentu, 1809.

Mengel, R. M. "John James Audubon." In *Dictionary of Scientific Biography*. New York: Charles Scribner's Sons, 1970.

Pearson, H. *Doctor Darwin*. London: J. M. Dent, 1930.

Peattie, D. C., ed. *Audubon's America*. Boston: Houghton Mifflin, 1940.

Personal communications to G. Weissmann from Mr. H. G. Button, archivist of Christ's College Cambridge, via Sir Hans Komberg.

Rafinesque, C. S. *A Life of Travels in North America and South Europe*. Philadelphia: F. Turner, 1836.

The Royal Society Book of Signatures 1660–1979. London: Royal Society, 1980.

INFLAMMATION AS CULTURAL HISTORY (1981/1998)

Baudelaire, C. *Les Fleurs du Mal* (bilingual ed.), trans. W. H. Crosby, p. 260. Brockport, N.Y.: Boa Editions, 1991.

Bernard, C. *Leçon sur les effets des substances toxiques et médicamenteuses*. Paris: Baillière, 1857.

Bowle, J. *The Imperial Achievement*. Boston: Little, Brown, 1974.

Ford, F. M. *Parade's End*. New York: Alfred A. Knopf, 1961 (orig. 1922).

Gide, A. *Journals*, Vol. IV, 1939–1949, trans. J. O'Brien. New York: Alfred A. Knopf, 1951.

Metchnikoff, E. *Immunity in Infective Diseases*. New York: Johnson Reprint Corp., 1968 (orig. 1905).

Monod, J. *Le hasard et la nécessité; essai sur la philosophie naturelle de la biologie moderne*. Paris: Éditions du Seuil, 1970.

Smith, L. R., K. L. Bost, and J. E. Blalock. "Generation of idiotypic and anti-idiotypic antibodies by immunization with peptides encoded by complementary RNA: a possible molecular basis for the network theory." *Journal of Immunology*, 138:7–9, 1987.

Thomas, H. *The Spanish Civil War*, p. 328. New York: Harper & Row, 1961.

Thomas, L. In: *Immunopathology of Inflammation*, ed. B. K. Forscher and J. C. Houck, p. 2. Amsterdam: Excerpta Medica, 1971.

GERTRUDE STEIN AND THE CTENOPHORE (1989/1998)

Brinnin, J. M. *The Third Rose*. Boston: Addison & Wesley, 1959.

Burns, E. M., and U. E. Dydo. *The Letters of Gertrude Stein and Thornton Wilder*. New Haven: Yale University Press, 1996.

Curtis, W. C. "Good old summer times at the MBL." *The Falmouth Enterprise*, Falmouth, Mass., August 12, 1955.

Freud, S., and J. Breuer. *Studies in Hysteria*. Reprint of the 1895 edition in *The Standard Edition of the Complete Psychological Works*, eds. J. Strachey and A. Freud. London: Hogarth Press, 1956.

James, W. *Essays on Faith, Ethics and Morals*. New York: New American Library, 1974.

Lewis, S. *Arrowsmith*. New York: Harcourt Brace, 1924.

Loeb, J. *The Organism as a Whole: From a Physico-Chemical Viewpoint*. New York: G. P. Putnam's Sons, 1916.
———. *The Mechanistic Conception of Life*, reprint ed. Cambridge, Mass.: Belknap Press of Harvard University, 1965.
McGrath, W. J. "Peter Gay: Freud." *New York Review of Books*, August 12, 1988.
Pauly, J. P. *Controlling Life: Jacques Loeb and the Engineering Ideal in Biology*. Oxford: Oxford University Press, 1987.
Skinner, B. F. "Has Gertrude Stein a secret?" *The Atlantic*, April 1935.
Solomons, L. M., and G. Stein. "Normal motor automatism." *Psychological Review* (Harvard Psychological Laboratory), 2:492–512, 1896.
Stein, G. *How to Write*. Paris: Plain Éditions, 1931.
———. *The Autobiography of Alice B. Toklas*. New York: Harcourt Brace, 1933.
———. *Lectures in America*, intr. W. Steiner. Reprint, Beacon Press, Boston, 1985; originally published by Random House, New York.
———. *Picasso*. London: B.T. Batsford, 1938.
———. *Selected Writings*. New York: Vintage Books, 1972.
———. *Paris: France*. New York: Liveright, 1970.
Steiner, W. *Exact Resemblance to Exact Resemblance. The Literary Portraiture of Gertrude Stein*. New Haven: Yale University Press, 1978.

PUERPERAL PRIORITY (1997)

Anon. History of University of Pennsylvania Department of Obstetrics and Gynecology via http://www.hupen.obgyn.edu.
Barker, F. *New-York Journal of Medical and Collateral Sciences, 3d series*, 3:105–107, 348–355, 1857.
Carter, K. C., and B. R. Carter. *Childbed Fever: A Scientific Biography of Ignaz Semmelweis*, p. 51. Westport, Conn.: Greenwood Press, 1994.
Charcot, J. M. "Pasteur." *Cosmopolitan (USA)*, 18:15–17, 1895.
Clarke, A. *New-York Journal of Medical and Collateral Sciences, 3d series*, 3:105–107, 1857.
Crosette, B. "Tally of world tragedy: women who die giving life." *The New York Times*, June 11, 1996.
Cullingworth, C. J. "Oliver Wendell Holmes and puerperal fever." *British Medical Journal*, 2:1161–1167, 1905.
"E. H." (Elisha Harris). "Review of Meigs, C. Childbed fevers." *New-York Journal of Medical and Collateral Sciences, New Series*, 15:258–260, 1855.
Gordon, A. *A treatise on the epidemic of puerperal fever of Aberdeen*. London: G. G. and J. Robinson, 1795.
Holmes, O. W. "The contagiousness of puerperal fever." *New England and Quarterly Journal of Medical Surgery*, 1:503–530, 1842.
———. *Puerperal Fever as a Private Pestilence*. Boston: Ticknor & Fields, 1855.
———. "For a Meeting of the National Sanitary Association, 1860." In *Collected Poems*, Vol. 2, p. 270. Boston: Houghton Mifflin, 1892.

———. Quoted in Osler, W. "Oliver Wendell Holmes." *Bulletin of the Johns Hopkins Hospital*, 42:86–89, 1894.

Horton, R. "Truth and heresy about AIDS." *The New York Review of Books*, 43:14–20, 1996.

"I.Z." (Zoltán, I.). "Semmelweis." In *Encyclopaedia Britannica* (1995), CD-Rom 2.0.

Lea, A. Quoted in Loudon, I. "Puerperal fever, the streptococcus, and the sulfonamides, 1911–1945." *British Medical Journal*, 295:485–490, 1987.

Lister, J. "On the antiseptic principle in the practice of surgery." *Lancet*, 2:353–356, 668–669, 1867.

———. "On the effects of the antiseptic system of treatment upon the salubrity of a surgical hospital." *Lancet*, 1:4–6, 40–42, 1870.

Louis, P. C. A. *Recherches sur les effets de la saignée dans quelques maladies inflammatoires, etc.* Paris: Baillière, 1835.

Mayrhofer, C. "Zur Frage nach der Ätiologie der Puerperalprocesse." *Monatsschrift für Geburtshilfe Frauenkrankherten*, 25:112–134, 1865.

Morton, L. T. *A Medical Bibliography*, 4th ed., p. 844. Aldershot, Hampshire: Gower, 1983.

Nightingale, F. via http://www.kumc.edu/service/clendening/florence.

Osler, W. "Oliver Wendell Holmes." *Bulletin of the Johns Hopkins Hospital*, 42:86–88, 1894.

Pasteur, L. "Septicémie puerpérale." *Bulletin de l'Académie de Médecine* (Paris) 2me séries, 8:505–508, 1879.

Rosenberg, C. E. *The Care of Strangers: The Rise of America's Hospital System*, pp. 132–133. New York: Basic Books, 1987.

Semmelweis, I. F. "Höchst wichtige Erfahrung über die Ätiologie in der Gebäranstalten epidemischen Puerperalfieber." *Zeitschrift Kaiserische & Königliche Gesellschafft der Aertze Wien*, 4(pt 2):242–244, 1847/8; 5:64–65, 1849.

———. *Die Ätiologie, der Begriff und die Prophylaxis der Kindbettfiebers.* Leipzig: CA Hartleben, 1861.

Vallery-Radot, R. T. *The Life of Pasteur*, trans. R. L. Devonshire, p. 289. Garden City, NY: Doubleday, Page & Company, 1926.

Weissmann, G. "Doctor Doyle and the case of the guilty gene." In *Democracy and DNA*, pp. 179–192. New York: Hill & Wang, 1996.

BÊTE NOIRE (1996)

Barker, P. *Regeneration*. New York: Dutton, 1992.

———. *The Eye in the Door*. New York: Dutton, 1994.

———. *The Ghost Road*. New York: Dutton, 1996.

Cannon, W. B. "The emergency functions of the adrenal medulla in pain and the major emotions." *American Journal of Physiology*, 33:356–372, 1914.

Douglas, K. "Bête Noire" (1944). In *Selected Poems*. London: Faber & Faber, 1964.

———. *Alamein to Zem Zem*. London: Faber & Faber, 1946.

Hippocrates. *The Book of Prognostics*. In *The Genuine Works of Hippocrates*, trans. F. Adams. Baltimore: William & Wilkins, 1958.

Kipling, R. In *Barrack Room Ballads* (1890). New York: Doubleday, Page & Company 1916.

Rivers, W. H. R. "Freud's theory of the unconscious." *Lancet*, 1:912–914, 1917.

Selye, H. A. "A syndrome produced by diverse nocuous agents." *Nature*, 138:32, 1936.

Thomas, L. *Late Night Thoughts on Listening to Mahler's Ninth Symphony*, p. 34. New York: Viking, 1983.

Young, A. *The Harmony of Illusions: Inventing Post-Traumatic Stress Disorder*. Princeton: Princeton University Press, 1995.

SCIENCE FICTIONS (1993)

Dionysius Longinus. *de Sublimi*. Venice: Bodoni, 1793 (A. Edelson library).

Eisenberg, D. *Encounters with Qi: Exploring Chinese Medicine* (with T. L. Wright). New York: Norton, 1985.

Gilbert, S. "Cellular politics: Ernest Everett Just, Richard B. Goldschmidt and the attempt to reconcile embryology and genetics." In *The American Development of Biology*, ed. R. Rainger, K.R. Benson, and J. Maienschein, Philadelphia: University of Pennsylvania Press, 1988.

Keller, E. F. *A Feeling for the Organism: The Life and Work of Barbara McClintock*. San Francisco: Freeman, 1983.

———. *Reflections on Gender and Science*. New Haven: Yale University Press, 1985.

La Follette, M. *Stealing into Print*. Berkeley: University of California Press, 1992.

Locke, D. *Science as Writing*. New Haven: Yale University Press, 1992.

Polanyi, K. *The Great Transformation*. New York: Octagon Books, 1975.

Rothfield, L. *Vital Signs*. Princeton: Princeton University Press, 1992.

Snow, C. P. *The Affair*. New York: Charles Scribner's Sons, 1960.

CALL ME MADAME (1996)

Bensaude-Vincent, B. *Langevin, 1872–1946: science et vigilance*. Paris: Belin, 1987.

Curie, E. *Madame Curie: A Biography*, trans. V. Sheehan. New York: Da Capo Press, 1986 (orig. 1937).

Curie, M. *Correspondance; choix de lettres, 1905–1934 [de] Marie [et] Irène Curie*. Paris: Éditeurs français réunis, 1974.

Giroud, F. *Marie Curie, a Life* (orig. *Femme honorable*, trans. L. Davis). New York: Holmes & Meier, 1986.

Igot, Y. *Monsieur et Madame Curie*. Paris: Didier, 1960.

Langevin, P. *La pensée et l'action*, ed. P. Laberenne. Paris: Editions sociales, 1964.

Pais, A. *Subtle is the Lord: The Science and the Life of Albert Einstein*. Oxford: Oxford University Press, 1982.

———. *A Tale of Two Continents: The Life of a Physicist in a Turbulent World*. Princeton: Princeton University Press, 1997.

Quinn, S. *Marie Curie: A Life*. New York: Simon & Schuster, 1995.

Woolf, H. ed. *Some Strangeness in the Proportion: A Centennial Symposium to Celebrate the Achievements of Albert Einstein.* Reading, MA: Addison–Wesley, 1980.

THE WOODS HOLE CANTATA (1984)

Duffus, R. L. "Jacques Loeb." *Century Magazine*, Vol. 2, pp. 374–385, 1925.
Loeb, J. *The Organism as a Whole: From a Physico-Chemical Viewpoint.* New York: G. P. Putnam's Sons, 1916.
———. "Biology and war." *Science*, 45:73–76, 1917.
———. *The Mechanistic Conception of Life*, reprint ed. Cambridge, MA: Belknap Press of Harvard University, 1965.
Osler, W. *The Old Humanities and the New Science*, intr. H. Cushing. Boston: Houghton Mifflin, 1920.
Osterhout, W. J. V. "Jacques Loeb." *Journal of General Physiology*, 8:9–42, 1928.

FOUCAULT AND THE BAG LADY (1982)

Foucault, M. *Madness and Civilization.* New York: Random House, 1965.
———. *Birth of the Clinic: an Archaeology of Medical Perception.* New York: Pantheon, 1973.
Laing, R. D. *The Divided Self: A Study of Sanity and Madness.* London: Tavistock Publishers, 1960.
Pinel, P. *A Treatise on Insanity*, 1806, trans. D. D. Davis. New York: Hafner Publishing Company, 1962.

IN QUEST OF FLECK:
SCIENCE FROM THE HOLOCAUST (1980)

Fleck, L. *Genesis and Development of a Scientific Fact: Introduction to the Study of Thoughtstyle and Thoughtcollective*, eds. T. J. Trenn and R. Merton. Chicago: University of Chicago Press, 1979 (first German edition, 1935).
———. "Specific antigenic substances in the urine of typhus patients." *Texas Reports on Biology & Medicine*, 9:697–708, 1947.
Fleck, L., and Z. Murczynska. "Leukergy." *Texas Reports on Biology & Medicine*, 9:709–734, 1947.
Korchak, H. M., K. Vienne, L. E. Rutherford, and G. Weissmann. "Neutrophil stimulation: receptor, membrane, and metabolic events." *Federation Proceedings*, 43:2749–2754, 1984.

AUDEN AND THE LIPOSOME (1981)

Auden, W. H. *Collected Poems of W. H. Auden.* New York: Random House, 1945.
———. *The Dyer's Hand.* London: Faber & Faber, 1963.
———. *About the House.* London: Faber & Faber, 1965.

———. In *The Place of Value in a World of Facts: 14th Nobel Symposium*, eds. A.W.K. Tiselius and S. Nilsson. New York: Wiley–Interscience, 1970.
Bangham, A. D. *Liposome Letters*. New York: Academic Press, 1983.
Bangham, A. D., M. M. Standish, and G. Weissmann. "The actions of steroids and streptolysin S on the permeability of phospholipid structures to cations." *Journal of Molecular Biology*, 13:253–259, 1965.
Carpenter, H. W. H. *Auden: a Biography*. Boston: Houghton Mifflin, 1981.
Connot, R. E. *Justice at Nuremberg*. New York: Carroll & Graf, 1983.
Diderot, D. *Rameau's Nephew*, trans. J. Barzun and R. U. Bowen. New York: Doubleday/Anchor, 1956.
Forster, E. M. *Two Cheers for Democracy*. New York: Harcourt Brace, 1951.
Hilberg, R. *Documents of Destruction*. Chicago: Quadrangle Books, 1971.
Mitscherlich, A., and F. Mielke. *Doctors of Infamy*. New York: H. Schuman, 1949.

NO IDEAS BUT IN THINGS (1985)

Coles, R., ed. *William Carlos Williams; The Doctor Stories*. New York: New Directions, 1984.
Mariani, P. *William Carlos Williams; A New World Naked*. New York: McGraw–Hill, 1981.
Ober, W. B. "William Carlos Williams: the influence of medical practice." *Journal of the Medical Society of New Jersey*, WCW Memorial Issue, 80:34–37, 1983.
Williams, W. C. *Collected Early Poems*. New York: New Directions, 1950.
———. *Make Light of It: The Collected Short Stories*. New York: Random House, 1950.
———. *The Autobiography of William Carlos Williams*. New York: New Directions, 1951.
———. *Collected Later Poems*. New York: New Directions, 1951.
———. *Paterson. Books 1–6*. New York: New Directions, 1963.

NOBEL WEEK 1982 (1983)

Nobel Foundation: Text of citations for the 1982 awards, 23 pp.

THEY ALL LAUGHED AT CHRISTOPHER COLUMBUS (1986)

Calin, A., and J. F. Fries. "An 'experimental' epidemic of Reiter's syndrome revisited. Followup evidence on genetic and environmental factors." *Annals of Internal Medicine*, 84:546–566, 1976.
Fiessinger, N., and M. E. Le Roy. "Contribution a' l'étude d'une épidémie de dysenterie dans la Somme." *Bulletin et Mémoires de la Société de Médecins Hôpital de Paris*, 40:2030–2069, 1917.
Granzotto, G. *Christopher Columbus: The Dream and the Obsession*. Garden City, NY: Doubleday, 1985.

Jones, E. *The Life and Work of Sigmund Freud*. Vol. 2: *The Years of Maturity 1901–1919*. New York: Basic Books, 1955.

Morison, S. E. *Admiral of the Ocean Sea: A Life of Columbus*. Boston: Little, Brown, 1942.

———. *Christopher Columbus, Mariner*. New York: Meridien/Viking, 1983.

Reiter, H. "Über eine bisher unerkannte Spirochaeteninfectim." *Deutsche Medizinische Wochenschrift*, 42:1535–1536, 1916.

Rolf, H. R. "An 'experimental' epidemic of Reiter's syndrome." *Journal of the American Medical Association*, 197:693–698, 1966.

Schoenrich, O. *The Legacy of Christopher Columbus*. Glendale, CA: Arthur H. Clark, 1949.

Williams, W. C. *In the American Grain*. New York: New Directions, 1925.

WESTWARD THE COURSE OF EMPIRE (1986)

Doctorow, E. L. *World's Fair*. New York: Random House, 1985.

Gidieon, S. *Space, Time and Architecture*. Cambridge, MA: Harvard University Press, 1952.

Jacobs, J. *The Death and Life of Great American Cities*. New York: Vintage/Random House, 1961.

Le Corbusier. *City of Tomorrow*. Cambridge, MA: MIT Press, 1971.

Manser, T., L. J. Wysocki, T. Gridley, R. I. Near, and M. L. Gefter. "The molecular evolution of the immune response." *Immunology Today*, 6:1–7, 1985.

Mumford, L. *The Culture of Cities*. New York: Harcourt Brace Jovanovich, 1970.

The Westsider, February 6, 1986.

Whitman, W. "Manahatta." In *Leaves of Grass*, ed. S. Bradley. 3 vols. New York: New York University Press, 1980.

NULLIUS IN VERBA: LUPUS AT THE ROYAL SOCIETY (1985)

Hill, C. *The Century of Revolution: 1603–1714*. New York: Norton, 1966.

———. *God's Englishman*. New York: Harper/Torchbooks, 1970.

Hooke, R. *Micrographia or Some Physiological Descriptions of Minute Bodies Made by Magnifying Glasses with Observations and Inquiries Thereupon*. ed. R.T. Gunther. New York: Dover Reprint Editions, 1938.

Merton, R. K. *On the Shoulders of Giants: A Shandean Postscript* (vicennial edition). New York: Harcourt Brace Jovanovich, 1985.

Ogg, D. *England in the Reign of Charles II*. Oxford: Oxford University Press, 1967.

Reinherz, E., D. Hans, T. Royer, J. Campen, R. Punia, F. Marnia, and A. Orespe. "The ontogeny, structure and function of the human T-cell receptor for antigen and major histocompatibility complex." *Biochemical Society Symposia*, 51:211–232, 1986.

The Royal Society: A Brief Guide to Its Activities. London: The Royal Society, 1981.

Stone, L. *The Causes of the English Revolution*. New York: Harper/Torchbooks, 1972.

Vitetta, E., and J. Uhr. "Immunotoxins." *Annual Review of Immunology*, 3:197–212, 1985.
Walzer, M. *The Revolution of the Saints: A Study in the Origins of Radical Politics.* New York: Atheneum, 1974.
Wiley, B. *The Seventeenth-Century Background.* Garden City, NY: Doubleday/Anchor, 1953.
Wong, W. W., L. Klickstein, J. A. Smith, J. H. Weiss, and D. T. Fearon. "Identification of a partial cDNA clone for the human receptor for complement fragments C3b/C4b." *Proceedings of the National Academy of Sciences (USA)*, 82:7711–7715, 1985.

SPRINGTIME FOR PERNKOPF (1985)

Conot, R. E. *Justice at Nuremberg.* New York: Carroll & Graf, 1984.
"The fate of Austrian scientists." *Journal of the American Medical Association*, 111:1778, 1938.
Hilberg, R., ed. *Documents of Destruction.* Chicago: Quadrangle Books, 1971.
Lancet. Vols. 1, 2 (1938).
Shirer, W. L. *The Nightmare Years.* Boston: Little, Brown, 1984.
Waugh, E. *When the Going Was Good.* Boston: Little, Brown, 1984.
Wiener Klinische Wochenschrift. Vol. 51 (1938).

THE BARON OF BELLEVUE (1986)

Asher, R. "Münchhausen syndrome." *Lancet*, I:339–341, 1951.
Barrter, F. C., P. Pronove, J. R. Gill, and R. C. McArdle. "Hyperplasia of the juxtaglomerular complex with hyperaldosteronism and hypokalemic alkalosis." *American Journal of Medicine*, 33:811–828, 1962.
Caswell, J. *The Romantic Rogue: Being the Singular Life and Adventures of Rudolph Erich Raspe, Creator of Baron Münchhausen.* New York: Dutton, 1979.
Honor, H. *Neo-classicism.* New York: Penguin, 1977.
Meadow, R. M. "Münchhausen syndrome by proxy." *Lancet*, 2:343–345, 1977.
Verity, C. M., C. Winckworth, D. Burman, D. Stevens, and R. J. White. "Polle syndrome." *British Medical Journal* 2:422–423, 1979.

THE DOCTOR WITH TWO HEADS (1988)

Baldick, R. *The First Bohemian: The Life of Henry Murger.* London: Hamish Hamilton, 1961.
Cantor, N. F. *Twentieth Century Culture.* New York: Peter Lang, 1988.
Collins, L., and D. La Pierre. *Is Paris Burning?* New York: Simon & Schuster, 1965.
Combat (Paris), no. 3 (May 1942).
Combat médical (Paris), no. 2 (March 1944).
Frizell, B. *Ten Days in August.* New York: Simon & Schuster, 1956.

Garnot, N. S. F. "L'Architecture hôpitalier au XIX siècle: L'Exemple parisien." *Les Dossiers du Musée d'Orsay*, no. 27 (1988).

Gramont, S. de. *The French.* New York: G. P. Putnam's Sons, 1969.

Guerard, A. *France: A Modern History.* Ann Arbor: University of Michigan Press, 1959.

Hillairet, J., and P. Payen-Appenzeller. *Dictionnaire historique des rues de Paris.* Paris: Les Éditions de Minuit, 1972.

Magraw, R. *France 1815–1914: The Bourgeois Century.* London: Fontana, 1983.

Marx, K., and F. Engels. *Basic Writings on Politics and Philosophy*, ed. L. Feuer. New York: Doubleday/Anchor, 1959.

Monod, R. *Les Heures décisives de la libération de Paris.* Paris: Éditions Gilbert, 1947.

Le Petit Journal (Paris; illustrated supplement), February 27, 1898.

Seigel, J. *Bohemian Paris.* New York: Elisabeth Sifton Books, 1986.

Simon-Dhouailly, N. *La Leçon de Charcot: Voyage dans une toile.* Paris: Musée de l'Assistance Publique, 1986.

———. *Musée de l'Assistance Publique de Paris* (catalogue). Paris: Musée de l'Assistance Publique, 1987.

Zeldin, T. *France 1848–1945* (3 vols.). Oxford: Oxford University Press, 1979.

Zola, É. *Le Docteur Pascal.* In *Oeuvres complètes.* Paris: Cercle du Livres Precieuses, 1960.

WORDSWORTH AT THE BARBICAN (1987)

Hartman, G. H. *Wordsworth's Poetry 1787–1814.* Cambridge, MA: Harvard University Press, 1971.

Nicolson, M. H. *Newton Demands the Muse.* Princeton: Princeton University Press, 1946.

Ozick, C. "Science and letters: God's work—and ours." *The New York Times Book Review*, September 27, 1987.

Vane, J. R. "Inhibition of prostaglandin synthesis as a mechanism of action for aspirin-like drugs." *Nature, New Biology*, 231:232–234, 1971.

Wordsworth, W. *Poems.* London: J. M. Dent, 1920.

LOSING A MASH (1987)

Hoffman, S. *Under the Ether Dome: One Doctor's Apprenticeship at Massachusetts General Hospital.* New York: Charles Scribner's Sons, 1987.

Klass, P. *A Not Entirely Benign Procedure: Four Years as a Medical Student.* New York: G. P. Putnam's Sons, 1987.

Konner, M. *Becoming a Doctor.* New York: Elisabeth Sifton Books, 1987.

Le Baron, C. *Gentle Vengeance: An Account of the First Years at Harvard Medical School.* New York: Richard Marek, 1981.

Orwell, G. *Homage to Catalonia.* London: Victor Gollancz, 1938.

TO THE NOBSKA LIGHTHOUSE (1990)

Bevilacqua, M. P., S. Stengelin, M. A. Gimbrone, and B. Seed. "Endothelial leukocyte adhesion molecule 1: an inducible receptor for neutrophils related to complement regulatory proteins and lectins." *Science*, 243:1160–1164, 1989.

Boutry, G. A. "Augustin Fresnel: his time, life and work." *Science Progress*, 36:587–604, 1948.

Bradford, W. *Of Plymouth Plantation 1620–1647*, ed. S. E. Morison. New York: Alfred A. Knopf, 1952.

Delbanco, A. *The Puritan Ordeal*. Cambridge, MA: Harvard University Press, 1989.

Marx, K., and F. Engels. *Basic Writings on Politics and Philosophy*, ed. L. Feuer. New York: Doubleday/Anchor, 1959.

Mencken, H. L. *Letters*, Selected and annotated by G. J. Forgue. New York: Alfred A. Knopf, 1961.

Morse, J. T. *Oliver Wendell Holmes: Life and Letters*, 2 vols. Boston: Houghton Mifflin, 1896.

Thomas, L. "The physiological disturbances produced by endotoxins." *Annual Reviews of Physiology*, 16:467–478, 1954.

———. *The Youngest Science: Notes of a Medicine Watcher*. New York: Viking Press, 1983.

Walzer, M. *The Revolution of the Saints: A Study in the Origins of Radical Politics*. Cambridge, MA: Harvard University Press, 1982.

Weissmann, G., and L. Thomas. "Studies on lysosomes I. The effects of endotoxin, endotoxin tolerance and cortisone on release of acid hydrolases from a granular fraction of rabbit liver." *Journal of Experimental Medicine*, 116:451–466, 1962.

Wood, G. S. "Struggle over the Puritans." *The New York Review of Books*, 36(17):26–34, 1989.

TITANIC AND LEVIATHAN (1989)

Ballard, R. *Discovery of the Titanic*. London: Hodder & Stoughton, 1987.

Bradford, W. *Of Plymouth Plantation 1620–1647*, ed. S. E. Morison. New York: Alfred A. Knopf, 1952.

Gay, P. *Sigmund Freud*. New York: Norton, 1988.

Marshall, L. *The Sinking of the Titanic and Great Sea Disasters*, ed. L. Marshall. Philadelphia: J. C. Winston, 1912.

Melville, H. *Moby-Dick; or, The Whale*. New York; Heritage, 1943.

Oceanus, 28 (Winter 1986–7; *Titanic* expedition issue).

Russell, T H. *Sinking of the Titanic*. New York: L. H. Walter, 1912.

Wade, W. C. *The Titanic*. New York: Penguin, 1986.

DAUMIER AND THE DEER TICK (1989)

Banville, T. de. *Petites études: Mes souvenirs*. Paris: G. Charpentier, 1882.

Baudelaire, C. *Oeuvres complètes*, ed. Y. G. Le Dantec and C. Pichois. Paris: Galimard, 1961.

———. *Les Fleurs du mal*, ed. J. Crépet, A. Blin, and C. Pichois. Paris: Galimard, 1968.

———. *Les Fleurs du mal*, trans. R. Howard. Boston: David Godine, 1982.

Bernard, C. *Introduction to the Study of Experimental Medicine*, trans. H. C. Green. New York: Schuman, 1949.

Brock, R. *Robert Koch*. New York: Springer, 1988.

Brookner, A. *The Genius of the Future*. London: Phaidon, 1971.

Burgdorfer, W., A. G. Barbour, S. F. Hayes, *et al.* "Lyme disease—a tick-borne spirochetosis?" *Science*, 216:1317–1319, 1982.

Clark, T J. *The Absolute Bourgeois*. London: Thames & Hudson, 1973.

———. *Image of the People: Gustave Courbet and the 1848 Revolution*. London: Thames & Hudson, 1973.

Faunce, S., and L. Nochlin. *Courbet Reconsidered*. Brooklyn, NY: The Brooklyn Museum, 1988.

Geoffroy, A. Quoted in P. Adhemar, *Daumier*. Paris: Pierre Tisne, 1954.

Goncourt, E. and J. *Gavarni*. Paris: Plon, 1954.

Lassaigne, D. *Daumier*, trans. E. B. Shaw. Paris: Hyperion, 1938.

Malawista, S. E., J. M. Oliver, and M. Rudolph. "Microtubules and cyclic nucleotides: on the order of things." *Journal of Cell Biology*, 77:881–886, 1978.

Malawista, S. E., A. C. Steere, and J. A. Hardin. "Lyme disease: a unique human model for an infectious etiology of rheumatic disease." *Yale Journal of Biology and Medicine*, 57:473–477, 1984.

Malawista, T. "Programme, février 1988." Paris: Privately printed, 1988.

Montgolfier, B. de. *Île St.-Louis* (catalogue). Paris: Musée Carnavalet, 1980.

Nochlin, L. "The development and nature of realism in the work of Gustave Courbet: a study of the style and its social and artistic background." Ph.D. dissertation, New York University, 1963.

———. *Realism*. New York: Penguin, 1971.

Rensberger, B. "A new type of arthritis found in Lyme, Conn." *The New York Times*, July 18, 1976.

Roueché, B. "The foulest and meanest creatures that be." *The New Yorker*, September 12, 1988.

Spielman, A. "Prospects for suppressing transmission of Lyme disease." *Annals of the New York Academy of Science*, 539:212–220, 1988.

———. "A changing landscape and the emergence of Lyme disease in North America," unpublished manuscript of talk at the Marine Biological Laboratory, Woods Hole, MA, July 1989.

Starkie, E. *Baudelaire*. London: Penguin, 1971.

Steere, A. C., S. E. Malawista, J. A. Hardin, *et al.* "Erythema chronicum migrans and Lyme arthritis: the changing clinical spectrum." *Annals of Internal Medicine*, 86:685–698, 1977.

———. "Lyme arthritis: an epidemic of oligoarticular arthritis in children and adults in three contiguous Connecticut communities." *Arthritis and Rheumatism*, 20:7–17, 1977.

Steere, A. C., T. F. Broderick, and S. E. Malawista. "Erythema chronicum migrans and Lyme arthritis: epidemiological evidence for a tick vector." *American Journal of Epidemiology*, 108:312–319, 1978.

Steere, A. C., M. S. Grodzicki, A. N. Kornblatt, *et al.* "The spirochetal etiology of Lyme disease." *The New England Journal of Medicine*, 308:733–740, 1983.

Telford, S. R., T. N. Mather, S. I. Moore, *et al.* "Incompetence of deer as reservoirs of the Lyme disease spirochete." *American Journal of Tropical Medicine*, 39:105–109, 1988.

Tocqueville, A. de. *Recollections*, trans. G. Lawrence. Garden City, NY: Doubleday, 1970.

Zurier, R. B., G. Weissmann, S. Hoffstein, *et al.* "Mechanisms of lysosomal enzyme release from human leukocytes II: effects of cAMP and cGMP, autonomic agonists and agents which affect microtubule function." *Journal of Clinical Investigation*, 53:297–309, 1974.

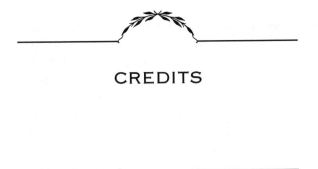

CREDITS

graphs *Gertrude Stein and Her Brother Finding Ctenophores on Quisset Beach, 1897,* and *Class Photo of the Marine Biological Laboratory Embryology Course, 1897;* the Trustees, The National Gallery, London, for J. Wright's *Experiments with the Air Pump;* the Collection, the Museum of Modern Art, New York, for Nadar's Woodburytype *Honoré Daumier* from Galerie Contemporaine; and the New-York Historical Society, New York, for J. J. Audubon's *Turkey Vulture (or Turkey Buzzard)* and Thomas Cole's *The Course of Empire* series.

INDEX

Note to the reader: The letters f and n are used to indicate that the entry is found in figures and notes, respectively.

Abortion
    and deaths of women, 255–267
    federal funding of, 260
Abyssinia, Mussolini's campaign in,
    192–193
Addams, Jane, 5
Adler, Felix, 277
Aiken, Conrad, 126
Alexander, Pamela, 23–24
Alzheimer's disease, medical model of, 102
Aniline dyes, Ehrlich's research on, 27–28
Anschluss, Hitler's campaign and, 193–196
Antigen, typhus, Fleck's research on,
    107–114
Anti-Semitism
    in Austria, 196–205
    failure of West to respond to, 203–205
Appell, Paul, 85
Arendt, Hannah, 200
Arthritis
    in Lyme disease, 301–304
    in Reiter's syndrome, 151–158

Asher, Richard, 210
Asylum
    absence of, 100–101
    defined, 99
    Foucault's critique of, 102–106
Auden, W. H., 78, 115–123, 116f
    political convictions of, 117–120
    on science, 117–123
Audubon, John James
    Darwin and, 9–24
    Turkey Vulture, 16f
Audubon, Lucy Bakewell, 18
Austria
    anti-Semitism in, 196–205
    post-Anschluss, 193–205
Autobiography of Alice B. Toklas, The, 38, 43
Automatic writing, Stein's work on,
    42–45

Babesiosis, diagnosis of, 306
Ballard, Robert, 271–272, 277
Baltimore, David, 72–73

Bangham, Alec, 117, 120
Barbican Arts and Conference Centre,
  235–243
Barker, Fordyce, 53–54
Barker, Pat, 62, 65–66, 67
Barrias, Ernest, 231, 233f
Bartter, Frederic, 208
Bartter's syndrome, 207–214
Barzan, Jacques, 4, 6
Baudelaire, Charles, 33–34, 297
  friendship with Daumier, 285–287
  on imagination, 303
  Les Fleur du Mal, 290
Bauer, Catherine, 161
Becquerel, Henri, 82
Benach, Jorge, 305
Bercovitch, Sacvan, 264
Berger, Bernard, 305
Bergström, Sune, 138–139, 144, 146
Bernard, Claude, 25–26, 85, 231, 297
  experimental medicine of, 77
Berra, Yogi, 169
Bevilacqua, Michael, 266
Bichat, Xavier, clinical medicine of, 77
Biological engineering, ethical issues in,
  120–123
Birds of America, The, 10, 13, 14, 17
Blake, William, 240
Boas, Ismar, 201
Bogart, Humphrey, 31, 33
Boissard de Boisdevier, Émile, 288
Bondi, Philip, 73
Borel, Emil, 85
Borrelia burgdorferi, identification of, 305
Boulton, Matthew, 18, 20
Bourgeoisie
  bohemian alter ego of, 226, 228
  in 19th-century France, 225–226
Bowle, John, 27
Boyle, Robert, 181
Bradford, William, 260–262, 271, 277
Bragg equation, 69, 71
  as value-free science, 76–77
Brain, W. Russel, 202
Braunwald, Eugene, 73
Breast cancer, prognosis for, 228–229
Brereton, John, 299–300
Breuer, Josef, 43
Bruant, Aristide, 230
Bryan, William Jennings, 276

Bubonic plague, 186
  victims of, 183
Buffon, Georges Louis Leclerc, Comte de,
  3–4, 10
Burgdorfer, Willy, 305, 306, 308
Burial at Ornans, The, 291, 292f
Bush, George, 260
Butt, Major, 274

Calin, A., 153
Calvin, John, 263
Cannon, Walter B., 66
Carnap, Rudolf, 110
Casablanca, 31–33
Cathartes aura, Audubon's description of,
  15, 16f
Céline, Louis-Ferdinand, 230
Cellular immunity, Metchnikoff's work on,
  26–28
Celsus, Cornelius, 25
Century of Revolution 1603-1714, The, 184
Challenger disaster, 277–278
Chamberlain, Neville, 193
Charles II, 179–180, 181, 185–186
Chauncy, Charles, 260–265
Chekhov, Anton, 2
Chiari, Dr. H., 194–195
Chicotot, Georges, 218–231
  as bicéphale, 220–221, 225–226
Childbed fever: See Puerperal sepsis
Cholera epidemics, in 19th-century Paris,
  291, 293–294
Churchill, Winston, 193
Cities
  Decentrist view of, 161–162
  generator of diversity in, 162–177
  immigrant renewal of, 176–177
  and self-destruction of diversity,
    165–167
  trends versus destiny in, 159–177
Citron, Julius, 111
Civil liberties, mental illness and, 101
Clark, T. J., 291, 293
Clarke, Alonzo, 54
Clemenceau, Georges, 229–230
Cognition, as social activity, 110–111
Cole, Thomas, 169, 170f–174f, 175–176
Coleridge, Samuel Taylor, 239
Coles, Robert, 125
Columbus, Bartholomew, 154

Columbus, Christopher, 149–158
  last voyage of, 149–151
  legacy of, 157
  second voyage of, 154
  as victim of Reiter's syndrome, 151–158
Complement receptors, abnormalities of, in
  systemic lupus erythematosus, 180–181
Connolly, Cyril, 117
Cotran, Ramzi, 266
Courbet, Gustave, 222, 223f, 288, 291, 292f
Course of Empire, The, 169, 170f–174f, 175–176
Crick, Francis, 75–76, 90n
Cromwell, Oliver, 180, 183–184
Cultural history, inflammation as, 25–48
Culture of Cities, The, 161
Curie, Eve, 83
Curie, Jacques, 82
Curie, Marie, 81–85
  Nobel prize of, 82
  personal history of, 84
Curie, Pierre, 81–85
  Nobel prize of, 82–83
  personal history of, 84
Curtis, W. C., 35, 37
Cuvier, Georges, 4, 10

Daladier, Edouard, 193
Dale, Sir Henry, 29, 139
Danforth, William, 276–277
Darwin, Charles, 14
  Audubon's influence on, 9–24
Darwin, Erasmus, 13, 14, 18–19, 20–21
Darwin, Robert, 14
Daudet, Leon, 84
Daumier, Honoré, 283–298, 286f
  friendship with Baudelaire, 285–287
  painting of, 294–297
  Paris of, 290–291
  political satire of, 284–285
  Ratapoil, 288, 289f, 290
  Washerwoman, 295, 296f
Davis, Elizabeth, 87
Day, Mary T., 203
de Gaulle, Charles, 33
Death and Life of Great American Cities, The,
  160
Decentrists, 161–162
Deinstitutionalization, 99–100
deKruif, Paul, 95
Delbanco, Andrew, 264

Delvaux, Alfred, 228
Derrida, Jacques, 33
Descent of Man and Selection in Relation to
  Sex, The, 13, 22–23
Diagnostic and Statistical Manual of Mental
  Disorders III, posttraumatic stress
  disorder in, 67
Diderot, Denis, 41, 91–93, 120–121
Dingell, John, 72
Disease
  social construction of, 68
  social distribution of, 183
DNA, as value-free science, 75–77
Doctor Stories, The, 128
Doctorow, E. L., 161
Doctor–patient relationship, Foucault's
  critique of, 104–105
Douglas, Keith, 61–62
Dreyfus, Alfred, 83
  vindication of, 230–231
Dreyfusards, political convictions of,
  229–230
Dryden, John, 182, 185
Dupee, Frederick, 37
Duval, Jeanne, 287
Dysentery, in Reiter's syndrome, 151–154

Ehrlich, Paul, 26, 188
Eliot, George, 4
Eliot, T. S., 130
Endean, Frederick C., 203
Endotoxin, in Shwartzman phenomenon,
  265–266
Engels, Friedrich, 263
England, Restoration: See Restoration
  England
Epinephrine, in necrosis, 265–267
Eppinger, Dr. M., 194, 200, 206
Erichsen, John, 64–65
Erythema chronicum migrans
  in Lyme disease, 302–304
  neurological symptoms with, 305
Essay, form and function of, 4–7
Ethics
  as genetically determined, 90–91
  lack of, in scientific applications, 118–119
  of liposomes, 120–122
Evelyn, John, 182
Evolutionary theory, French version of, 4

Experimentation, testing of truth by, 179–189

Fabians, 4–5
Fascism: see also National Socialism; Nazism
    as cultural inflammation, 29–31
Faure, J. L., 225
Faÿ, Bernard, 46–47
Fearon, Douglas, 180–181
Feder, Ned, 74
Felig, Philip, 73
Feminism, on construction of science, 70–71
Fiction, science as, 71–75
First Trial of X-ray Therapy for Cancer of the
    Breast, The, 218, 219f, 220–222
Fleck, Ludwik, 75, 107–114
Flight-or-fight response, 66
Ford, Ford Madox, 29
Form
    essay and, 4–7
    science and, 3–4
Forster, E. M., 122
Foucault, Michel, 33
    mental illness critique of, 97–106
    on mental institutions, 102–106
France
    bourgeois century in, 226
    under Nazi occupation, 216–217
France, Anatole, 26
Franco, Francisco, 193
Frankl, Oskar, 201
Franklin, Benjamin, 2
French Resistance, 231
    monument to, 217
Fresnel, Augustin-Jean, 256
Freud, Sigmund, 43, 275
    on traumatic memory, 65
Fries, J. F., 153
Frigate bird, in theory of natural selection,
    10–13, 22
Fuller, Margaret, 7
Fully, George, 217–218

Gallo, Robert, 73
García Márquez, Gabriel, 137–138, 140, 142,
    147
Gautier, Théophile, 288
Gender, science and, 70–71
Genesis and Development of a Scientific Fact,
    109–114

Genetics, sociobiological perspective on,
    119–120
Geoffroy, Gustave, 290–291
Gervex, Henri, 225, 226, 227f
Gide, André, 30–31
Gidieon, Siegfried, 161
Gimbrone, Michael, 266
Goddard, Jonathan, 184
Golot, Édmond, 226
Good, Robert, 73
Goodwin, Aubrey, 202–203
Gordon, Alexander, 50
Gosnold, Bartholomew, 299
Gould, Stephen Jay, 9
Granger, Thomas, 260, 265
Grant, Robert Lee, 14, 21
Grant, Ulysses S., 300
Granzotto, Gianni, 155
Graunt, John, 181, 183
Graves, Robert, 62, 66, 74
Grunwaldt, Edgar, 305
Guggenheim, Benjamin, 274

Haagen, Eugen, 118
Hall, Marshall, 65
Hardin, John, 304
Harmony of Illusions: Inventing
    Post-Traumatic Stress Disorder, The, 63–64
Harris, Dr. Elisha, 53
Harris, Mrs. Henry B., 274
Heidegger, Martin, 118
Henreid, Paul, 31–33
Henslow, John Stevens, 15, 17–18
Hill, Christopher, 184
Hinnant, Kathryn, 265
Hippocrates, 62
Histoire Naturelle, 10
Hitler, Adolf, 198
    Austrian medical establishment and, 195
    Austrian plebiscite and, 193–196
Hoffman, A., 200
Hollstrom, E., 305
Holmes, Oliver Wendell, 1, 2–3, 59, 263, 267
    on puerperal sepsis, 49–54
Homelessness, deinstitutionalization and,
    97–101
Homer, Winslow, 9
Hooke, Robert, 180, 181
Horder, Lord, 202
Howard, Ebenezer, 160–161

Hugo, Victor, 228
Human immunodeficiency virus epidemic,
    social context of, 308
Humoral immunity, Ehrlich's work on, 26
Hysteria
    Flaubert's treatment of, 77–78
    Solomons' and Stein's work on, 43–44

Imanishi-Kari, Dr., 72–73
Immigrants, urban renewal by, 176–177
Immunochemistry, origins of, 28
Immunology, after World War II, 1887
Inflammation
    as cultural history, 25–48
    Thomas on, 28–29
    white blood cells in, and Fleck's work
        on typhus vaccine, 108
Ionizing radiation, Curies' work on, 82
Isherwood, Christopher, 117
Ismay, J. Bruce, 273
Ixodes dammini
    in babesiosis, 306
    identification of, 304–305
    in Lyme disease, 298

Jacobs, Jane, 160–167
James, Alice, 65
James, Henry, 40, 65
James, William, 40–42, 45, 46, 65
Jameson, Robert, 14–15
Joliot-Curie, Frédéric, 83–84
Joliot-Curie, Irène, 83–84
Jones, Ernest, 275
Journal of the American Medical Association,
    on Jewish physicians in Austria, 200–201

Kardiner, Abram, 66
Keats, John, 239, 240
Keller, Evelyn Fox, 70–71
Kipling, Rudyard, 63
Kitchener, General, 27
Klug, Aaron, 141, 144
Knoepfelmacher, W., 201
Kornberg, Arthur, 90n
Kornberg, Hans, 74
Kraft, Joseph, 304
Kuhn, Thomas, 109–110

La Follette, Marcel C., 72
Laing, R. D., 101–102

Lamarck, Jean Baptiste, 4, 21–22
    "First Law" of, 11, 13
    "Second Law" of, 11
Lamb, Charles, 239, 242
Lancet, The, on Jewish physicians in Austria,
    202–206
Langevin, Paul, 81, 82, 83, 85
Leclerc, Georges Louis: See Buffon, Georges
    Louis Leclerc, Comte de
Leukergy, Fleck's work on, 108–109
Lewis, Wyndham, 42
Lifton, Robert Jay, 200
Lincoln Center, as neighborhood catalyst,
    164–165
Lipmann, Fritz, 73
Liposomes, discovery, uses, and ethical
    issues, 120–122
Lippmann, Gabriel, 85
Lister, Joseph, puerperal sepsis and, 57
Literature
    economic perspectives on, 78
    Freudian perspectives on, 77–78
    science as, 77–79
Locke, David, 75–76
Locke, John, 182
Loeb, Benedict, 91
Loeb, Jacques, 40–42, 71, 87–96
    as antispiritual champion, 89–90
    mechanistic philosophy of, 94–95
    on pseudobiology of race, 93
    on war, 92–94
Longinus, Dionysius, 69
Lorre, Peter, 31
Louis, Pierre C.-A., 50–51
Lyme disease, 283
    original diagnosis of, 301–302
    prevention of, 307
    research on, 298–309
    social context of, 300–309

MacGillivray, William, 15
Madness and Civilization, 102
Making of Americans, The, 44
Malawista, Stephen, 283
    Lyme disease research of, 298–309
Malawista, Tobé, 283–284, 288, 294, 295, 300,
    305
Mall, Franklin Pierce, 42
"Manahatta," 177
Manet, Edouard, 222, 224f

Manic-depressive psychosis, medical
   model of, 102
Mann, Erika, 115, 116f, 120, 123
Marshall, Logan, 274
Mather, Cotton, 260, 264
Mayer, Louis B., 5–6
McClintock, Barbara, 71
McFarland, Joseph, 26
McGrath, William J., 43
Meadow, Roy, 210
Mechanistic Conception of Life, 42
Medical model, Laingian assault on, 102
Medicine
   military parallels to, 245–247, 254
   under Nazism in Austria, 194–205
   self-destruction of diversity in, 167–169,
      175
Meigs, Charles, 53
Melville, Herman, 269–270, 280–281
Memory, traumatic, 65; *see also*
   Posttraumatic stress disorder
Mencken, H. L., 262
Menkin, Valy, 26
Mensch, Judy, 301
Mental illness
   civil rights issues and, 101
   Foucault's critique of, 97–106
   medical model of, 101–102
   Pinel's approach to, 103–106
   20th-century approaches to, 100
Mental patients, deinstitutionalization of,
   99–102
Mesmerism, commission on, 2–3
Metchnikoff, Élie, 26–27
Microscopy, development of, 186
Miller, Perry, 264
Millet, Jean François, 291
Molecular biology
   after World War II, 187–188
   cultural implications of, 31–33
Monoclonal antibodies, applications of,
   187–188
Monod, Jacques, 32
Morgan, J. P., 273
Morison, Samuel Eliot, 149–150, 152, 155,
   264
Mozart, Wolfgang Amadeus, 238
Müller, Hermann J., 90n
Mumford, Lewis, 160, 161
Münchausen's syndrome, 209–214

Münchhausen-by-proxy, 210–211
Murger, Henri, 226
Murray, Polly, 301
Mussolini, Benito, 193

National Socialism; *see also* Fascism; Nazism
   and Austrian medical establishment,
      193–205
   French origins of, 230
   Western medical establishment and,
      200–206
Natural selection
   Darwin's statement of, 12
   frigate bird in theory of, 10–13
Nazism; *see also* Fascism; National Socialism
   and abuse of science, 118–119
   as cultural inflammation, 29–31
   and Fleck's work on typhus vaccine,
      107–114
   French collaboration with, 216–217
   French Resistance and, 217–218, 231
Necrosis, epinephrine's role in, 265–267
Neighborhoods, revival of, 162–166
Neuroendocrinology, Cannon's
   contribution to, 66
Neurosis, traumatic, 66
Neutrophils, responses of, 109
Newton, Isaac, 180, 181
   Romantics' rejection of, 239
Nicolson, Marjorie Hope, 239–240
Nightingale, Florence, puerperal sepsis
   and, 57–58
Nobel, Alfred, 137
Nobel Prize awards, 137–148
   to Curies, 82–83
Nobl, Gabor, 201
Noer, H. Rolf, 152–153

Odocoileus virginianus, in Lyme disease, 298–
   300
*On the Origin of Species by Means of Natural
   Selection or the Preservation of Favoured
   Races in the Struggle for Life,* 10–12, 22
Ontogeny, color transitions as example of,
   12
Organism as a Whole, The, 41, 91
Ornithological Biography, 10, 12
Osler, William, 38, 48, 54
   on war, 92
Ovando, Don Nicolás de, 150–151

Ozick, Cynthia, 240

Painlevé, Paul, 85
Parthenogenesis, Loeb's predictions for, 90
Pascal, Gabriel, 6
Pasteur, Louis, puerperal sepsis and, 58–59
Paterson, 125, 129, 130–136
Pepys, Samuel, 182
Pernkopf, Dr. Edward, 194–205
Pétain, Maréchal, 37, 46–47
Petty, William, 181, 182, 184
Phagocytosis, Metchnikoff's work on, 26–28
Philosphie Zoologique, 11
Photosensitivity, Loeb's research on, 94
Phylogeny, Audubon and, 12
Physics, development of, 186
Picquart, Lt. Col., 230
Piezoelectricity, Curies' work on, 82
Pinel, Philippe, 231, 232f
  Foucault's critique of, 103–104
  on madness and its cure, 105–106
Polle's syndrome, 210–211
Polonium, Curies' discovery of, 83
Posttraumatic stress disorder, 62–68
  in DSM-III, 67
  social construction of, 63–64, 68
Pound, Ezra, Williams' friendship with, 127, 131
Powell, Anthony, 237
Priestley, Joseph, 18, 20
Proudhon, Pierre Joseph, 293
  French National Socialism and, 230
  on value of women, 222
Puerperal sepsis, 49–59
  conquerors of, 49
  deaths due to, 49–50
  Gordon's theory about, 50
  Holmes' theory about, 49–54, 267
  Lister's theory about, 57
  Nightingale and, 57–58
  Pasteur's theory about, 58–59
  Semmelweis' theory about, 54–57
Puritans
  sexuality and, 260–262
  traditional authority and, 263–264

Race
  National Socialist concept of, 197–200
  pseudobiology of, 93
Radium, Curies' discovery of, 83

Rafinesque, Constantine, 13
Railway spine, 65
Rains, Claude, 33
Ranzi, Egon, 200–201
Raspe, Rudolph Erich, 207, 212–213
Realism, in art of Daumier and Baudelaire, 297
Règne Animal, La, 10
Reichenbach, Hans, 109
Reinherz, Ellis, 73, 187
Reiter, Hans, 151
Reiter's syndrome
  Columbus as victim of, 151–158
  diagnosis of, 151–152
  genetic predisposition to, 153–154
  Noer's account of, 152–153
  transmission of, 155–156
Renan, Ernest, 25
Restoration England
  scientific discoveries in, 181, 186–187
  threats to health in, 183
Riefenstahl, Leni, 118
Rivers, W. H. R., 65–66
Rizak, E., 194, 200, 206
Robert-Fleury, Tony, 231, 232f
Roentgen, Wilhelm, 82
Rollender, A., 199
Romanticism, science and, 239–243
Roosevelt, Theodore, 300
Rostand, Jean, 25, 230
Rothfield, Lawrence, 77
Roueché, Berton, 305
Royal Society of London, 179–189
  early discussions of, 184–185
  intellectual roots of, 183–184

St. Hilaire, Geoffroy, 4, 10, 13
Samuelsson, Bengt, 139, 144, 147
Sassoon, Siegfried, 62, 66
Schachman, Howard, 74
Schizophrenia
  chronic, 97–100
  Laing's view of, 101–102
Schlick, Moritz, 110
Schuschnigg, Kurt von, 201
Science
  aesthetic component of, 3–4
  Auden's approach to, 117–123
  and commerce and trade, 186–187
  cultural history and, 25–48

Science (cont.)
curiosity versus utility as impetus for,
186–187
Dreyfusards and, 229–230
fictional, 71–75
versus fraud, 72–73
imagination in, 303–305, 308–309
as literature, 77–79
mistrust of, 122–123
Nazi abuse of, 118–119
in Restoration England, 186–187
Romantic revolution and, 239–243
self-cleansing aspect of, 74–75
social construction of, 70–71
social context of, 110–114
thought collective and, 110–114
20th-century reaction against, 2
value-free, 69–79
Nazi perspective on, 196–200
physical laws and, 71–72
Whig approach to, 4–5
as writing, 75–77
Scleroderma, 97–99
Scrimenti, R. J., 305
Sedgwick, Adam, 17
Seigel, Jerrold, 226
Selye, Hans, 66–67
Semmelweis, Ignaz Phillip, puerperal fever
and, 54–57
Sepsis, puerperal: See Puerperal sepsis
Shakespeare, William, on "PTSD," 62–63
Shaw, George Bernard, 4–6, 82
Shell shock: See Posttraumatic stress
disorder
Shigella flexneri, 151, 155; see also Reiter's
syndrome
Shwartzman phenomenon, 265–266
Silverman, Joseph, 277
Sjögren's syndrome, 98
Skinner, B. F., 40, 42–45
Sklodowska, Maria Salomea: See Curie,
Marie
Smith, Steven, 264–265
Smith, William Alden, 272
Snow, C. P., 69, 72
Snydman, David, 301–302
Sociobiology, political implications of,
119–120
Solomons, Leon M., 42–43
South, Robert, 184

Spender, Stephen, 117
Spielman, Andrew, 308
work on babesiosis, 306
Spray, Thomas, 186
Springer, Timothy, 266
Stalinism, as cultural inflammation,
29–31
Steere, Allen C., 302–304, 306, 308
Stein, Clarence, 161
Stein, Gertrude, 35–48
departure from medicine, 38
experiments with automatic writing,
42–45
fascist infatuation of, 37
research career of, 42–46
during World War II, 46–47
on writing, 45–46
Stein, Leo, 35, 37
Stewart, Walter W., 74
Stigler, George, 141, 145–147
Stimpson, Catherine, 37
Stone, Lawrence, 186–187
Stowe, Harriet Beecher, 263
Straus, Mrs. Isidor, 274
Stress, Selye's theory of, 66–67
Swainson, William, 17
Systemic lupus erythematosus
antibody production in, 182, 187
controversy over, 180–183, 188

Tender Buttons, 44
Theism, James' defense of, 41–42
Thomas, Lewis, 68, 258, 265–266
on inflammation, 28–29
Thomson, William (Lord Kelvin), 82
Thought collective, Fleck's concept of,
110–114
Three Lives, 38–39, 43
Three Sisters, 45
Titanic
moral lessons of, 273–275
sinking of, 272–273
steerage passengers of, 275–277
voyage to wreckage of, 278–280
Tocqueville, Alexis de, 291
Toklas, Alice B., 37
Traumatic memory, 64–65; see also
Posttraumatic stress disorder
Traumatic neurosis, 66
Trilling, Lionel, 4, 6

Tropisms
    instincts as, 90
    Loeb's descriptions of, 94
Trotter, Wilfred, 4–5
Turkey buzzard (*Cathartes aura*), Audubon's
    description of, 15, 16f
Typhus antigen, Fleck's research on, 107–114

Uhr, Jonathan, 188
Urban planning, perspectives on, 160–167
Urethritis, in Reiter's syndrome, 151,
    152–153
Uveitis, in Reiter's syndrome, 151, 152–153

Vaccine, typhus, Fleck's development of,
    107–114
Van Doren, Mark, 1–2, 7
van Eyck, Jan, 228
Van Vechten, Carl, 40
Vane, John, 139, 141, 144, 146
*Variation of Animals and Plants Under
    Domestication, The*, 13
Vienna Circle, 110
Virchow, Rudolf, 26
Vital Signs, 77
Vitetta, Ellen, 188
Von Euler, Ulf, 138–139

Wallis, John, 184
Walzer, Michael, 263
War, Loeb's responses to, 92–94
Washerwoman, The, 295, 296f
Wassermann reaction, social context in
    development of, 111–112
Waste Land, The, 130
Watson, James, 75–76, 90n
Watt, James, 18
Waugh, Evelyn, 191–192, 206
Webb, Beatrice, 5
Webb, Sidney, 5

Wedgwood, Josiah, 18
Weismann, August, 23
Wells, H. G., 5
Whewell, William, 17
Whigs, latter-day, 4–5
Whitehead, Alfred North, 6, 75
Whitman, Walt, 130, 177
Widener, Harry Elkins, 274
Wiener Klinische Wochenschrift, 194–200, 206
Wilde, Oscar, 2
Wilkins, John, 184
Williams, William Carlos, 125–136
    on analytic insights, 132–133
    on Columbus, 157
    European travel, 131
    family history, 130–131
    literary awards of, 135
    medical records of, 133–134
    objectivist, realistic mode of, 130
    on poetic line, 130
    on voices of patients, 135
Williams, William Eric, 133–134
Wilson, Edmund, 47, 126
Wilson, Kenneth G., 141, 144
Winthrop, John, 262–263
Wood, Gordon, 264
Wordsworth, William, 235–243
World War II, European boundaries during,
    193–194
Wren, Christopher, 182, 184, 235
Wright, Joseph, 20, 37
Wright, Samson, 203
Writing
    automatic, Stein's work on, 42–45
    science as, 75–77

Young, Allan, 63–64, 68

Zola, Émile, 228, 229, 294
Zoonomia, 18, 20–21, 22

ACKNOWLEDGMENTS

Many of these essays were first written for *Hospital Practice* and *MD* magazines, and I again acknowledge my debt to David Fisher, who presided over those monthly tributes to the two cultures. Among the several editors who have helped me since then, special thanks are due to Elisabeth Sifton as well as to my agent, Gloria Loomis. Those to whom I am newly indebted are the cadre of the New-York Historical Society who have been so helpful in my search for the several Darwin–Audubon connections. These include Stewart Desmond, who is responsible for the title of the title essay, Nancy Katzoff, who first asked me to speak at the Society about Audubon as scientist, as well as Wendy Shadwell, Curator of Prints, and May Stone—indeed, the entire staff of that splendid library on Central Park West. As with my earlier writing, three other libraries have been indispensable: those of the Marine Biological Laboratory–Woods Hole Oceanographic Institution in Woods Hole, the Villa Serbelloni (Rockefeller Foundation) in Bellagio, and the private library of Madame Arlette Gaillet in Paris. Sir Hans Kornberg, former master of Christ's College, Cambridge, was of immediate help with information on Darwin at Cambridge; Catherine Stimpson, of the MacArthur Foundation, corrected some of my views of Gertrude Stein; while Jean-Pierre Changeux, of the Collège de France, was kind enough to comment on the French text of "Inflammation as Cultural History." Versions of the uncollected essays have appeared elsewhere: "Puerperal Priority" in *Lancet*, "Bête Noire" in *The London Review of Books*, "Science Fictions" in *The Yale Review*, and "Call Me Madame" in *The New Criterion*; "Gertrude Stein and the Ctenophore" and "Inflammation as Cultural History" are reiterations of themes first developed in 1988 for the Centennials of the Marine Biological Laboratory and Johns Hopkins School of Medicine.

As always, my first reader, Ann Weissmann, made sure that the language of science did not interfere with the language of sense, and my astute editor Erika Goldman is to be thanked for rescuing my older essays from slow oxidation in library stacks. There is no end to the thanks I owe to Andrea Cody and Debbie Perez for their editorial and managerial help.

Finally, the subtitle of my book reflects the debt I owe to those who gave me a sense of the past in the middle of the twentieth century: Jacques Barzun, Joseph Wood Krutch, Mark Van Doren, and Lionel Trilling. Trilling set the tone in "The Liberal Imagination": "we would perhaps be stronger if we believed that Now contained all things, and that we in our barbarian moment were all that had ever been. Without a sense of the past we might be more certain, less weighted down and apprehensive. We might also be less generous, and certainly we would be less aware. In any case, we have a sense of the past and must live with it, and by it."

ABOUT THE AUTHOR

Gerald Weissmann is a professor of medicine and director of the Division of Rheumatology at New York University Medical Center. He was educated at Columbia College and at New York University, where he obtained his medical degree and did postdoctoral work in biochemistry with Severo Ochoa. He has also studied at the Strangeways Research Laboratory with Dame Honor Fell and at the Hôpital St. Antoine in Paris. He has received many awards, including the Lila Gruber Award for Cancer Research, a Guggenheim Fellowship, the Allesandro Robecchi Prize for Rheumatology, and the Distinguished Investigator Award of the American College of Rheumatology. His research work has focused on the cell biology of inflammation, with emphasis on lipid remodelling and lipid mediators. A past president of the American College of Rheumatology and the Harvey Society, Dr. Weissmann is editor-in-chief of *Inflammation*. His essays and reviews of cultural history have been published in *The New Republic*, *The New Criterion*, *the London Review of Books*, and The *New York Times Book Review* and have been collected in six volumes, from *The Woods Hole Cantata* (1985) to *Year of the Genome* (2001).